Exergetic Aspects of Renewable Energy Systems
Insights to Transportation and Energy Sector for Intelligent Communities

Evanthia A. Nanaki
Department of Mechanical Engineering
University of Western Macedonia
Kozani, Greece

George Xydis
Department of Business Development and Technology
Aarhus University, Herning
Denmark

CRC Press
Taylor & Francis Group
Boca Raton London New York

CRC Press is an imprint of the
Taylor & Francis Group, an **informa** business
A SCIENCE PUBLISHERS BOOK

Taylor & Francis Group
6000 Broken Sound Parkway NW, Suite 300
Boca Raton, FL 33487-2742

First issued in paperback 2020

ISBN-13: 978-1-138-08858-0 (hbk)
ISBN-13: 978-0-367-77935-1 (pbk)

Library of Congress Cataloging-in-Publication Data

Names: Nanaki, Evanthia A., author. | Xydis, George (Mechanical engineer), author.
Title: Exergetic aspects of renewable energy systems : insights to transportation and energy sector for intelligent communities / Evanthia A. Nanaki (Department of Mechanical Engineering, University of Western Macedonia, Kozani, Greece), George Xydis (Department of Business Development and Technology Aarhus University, Herning, Denmark).
Description: Boca Raton, FL : CRC Press, 2019. | "A science publishers book." | Includes bibliographical references and index.
Identifiers: LCCN 2019018767 | ISBN 9781138088580 (hardback)
Subjects: LCSH: Renewable energy sources. | Exergy. | Power resources.
Classification: LCC TJ808 .N36 2019 | DDC 333.79/401--dc23
LC record available at https://lccn.loc.gov/2019018767

Visit the Taylor & Francis Web site at
http://www.taylorandfrancis.com

and the CRC Press Web site at
http://www.crcpress.com

"Dedicated to my beloved parents & family, for their love, endless support & encouragement".

The best way to predict the future is to create it.

– Evanthia A. Nanaki

"To my wife and my daughters"

"Failure should be part of the curriculum in all levels of the education system"

– George Xydis

Preface

The transport sector is essential to all human activities as well as critical to social and economic development. Sustainable, secure and competitive energy supply and transport energy systems are at the heart of world strategies for a low carbon society. Innovation and deployment of new clean technologies are essential for a successful transition to a new sustainable and low carbon energy system. As the transport sector is responsible for about one quarter of global energy-related carbon emissions, it can be deduced that without aggressive and sustained policies, carbon dioxide emissions from this sector could double by 2050. Ambitious climate action plans can play a major role in denting the emissions curve and putting the world on track to reach full climate neutrality in the second half of the century. Therefore, it is imperative for transportation to be based on energy-efficient and low-emission modes that rely on clean energy. In order to address the abovementioned challenges, an integrated approach, encompassing technological, environmental as well as social aspects should be taken into consideration.

To achieve this goal, sustainable energy planning encompassing the concept of smart cities has a high potential to significantly contribute to the achievement of global energy and climate targets. For improved energy efficiency, it is essential to find low or zero carbon solutions for the urban environment. In smart cities, energy should also be used in a smart way, that is by reducing the energy degradation in terms of capacity to generate useful work. Technology is of great help in achieving this target, providing innovative and more efficient ways to respond to the increasing demand for more sophisticated and complex transport services. To exploit technological opportunities, city planners, administrators, citizens, entrepreneurs and all other stakeholders must reconsider the way they have approached transport and energy systems until now. Furthermore, national governments need to assess the long-term advantages and disadvantages of their available energy sources to reconcile the pressures induced by the move towards competitive markets with the requirements for a sustainable energy-strategy.

With these consideration in mind the authors of this book established an integrated approach against new challenges, assessed the weaknesses of prevailing theories, highlighted deficits in taking the exergy basics into account, and the need to understand global interactions within the process of energy transitions. Thus, this book is an attempt to capture the changing nature of complex energy systems by integrating and management of energy supply with predominant exploitation of local resources (e.g. waste heat, renewables) through the fundamental concept of exergy. For the first time, the concept of sustainable energy planning is correlated- via the method of exergy analysis- within the smart cities' perspective. The present contribution employs thermodynamic methods of analysis to evaluate the sustainability of energy systems.

This book proposes an analytical framework, based on exergy analysis as well as on Life Cycle Analysis, in order to support smart city planning, with the aim to provide decision makers with a useful tool to compare and understand the energy-smartness of different alternatives, and to address future energy urban policies.

The book illustrates the exergy concept in relation to the energy and transportation sector aiming at altering developers' thinking on increasing the efficiency of these sectors in relation to their surroundings. The basics of sustainable energy planning are presented, though however the fundamental concept of exergy applied to both above mentioned sectors is the heart of this book. The modern energy system under the smart cities' perspective is analytically presented. Transportation in urban environment and interaction with the intelligent energy systems is studied. The exergy principle is used within regional planning; referring to sustainable spatial design (improved use of energy flows, energy demand, e.g. next to industrial areas). Furthermore, possibilities through intelligent energy systems implemented were presented for energy conversion, distribution and storage to increase the integration of sustainable energy to the system and deliver energy to the end user under the most suitable scheme. An exergy planning approach is focusing in that direction and examines smart grids, electricity markets and how much they interact with each other.

The structure of this book is such that each chapter—describing a particular energy sector—is completely self-contained. The book is intended for anyone in need of comprehensive understanding of the key challenges of energy transition to zero carbon cities and the methods to provide solutions - in terms of system's optimization. Anyone would be able to reference a particular transport or energy efficiency option within the concept of smart or intelligent community city that is of interest without having to read other chapters.

Hopefully, this book will meet the needs of energy and transport policy planners and graduate students, economists, business, NGOs and researchers that deal with smart city technology, alternative fuels, renewable energy, and sustainable development. As the book attempts to fill the present knowledge gap by providing the necessary fundamentals, explaining the practical aspects of sustainable energy planning and spatial design, it can play a significant role in decision making, with regard to sustainable energy design both at a national and local level.

Evanthia A. Nanaki
George Xydis

Contents

Introduction

E. Nanaki[1] and G. Xydis[2]

[1] University of Western Macedonia, Department of Mechanical Engineering, Bakola & Sialvera, Kozani 50100, Greece
[2] Department of Business Development and Technology, Aarhus University, Birk Centerpark 15, 7400 Herning, Denmark

1.1 Energy, Climate Change, Smart Cities and Renewable Energy Integration

Energy is essential to all human activities as well as critical to social and economic development. Currently, the majority of energy systems are based on fossil fuel consumption leading to climate change and air pollution. Reducing energy use and increasing energy efficiency are crucial aspects for urban areas such as cities. Taking into consideration the fact that the worldwide population is growing, it is predicted that the fossil fuel consumption will keep on rising until 2050 [IEA energy outlook, 2017]. In this direction, many policies, such as the Paris Agreement, have been established. The Paris Agreement is characterized by a bottom-up approach to global cooperation, where each country delivers national inventories of greenhouse gases (NIR) and prepares Nationally Determined Contributions (NDC), which have to be adjusted and strengthened every five years [Doelle, 2016].

The Paris Agreement aims to keep global temperature rise this century below 2°C above pre-industrial levels and to pursue efforts to limit the temperature increase even further to 1.5°C. Nonetheless, in order to achieve 25% emissions reduction in 2040, opening the road to zero emissions in 2100, coordinated efforts of all stakeholders are imperative to take place. The introduction of renewable energy into the energy system of urban areas is a key element for the creation of smart and sustainable low carbon cities. In this direction, sustainable energy planning encompassing the concept of smart city has a high potential to contribute to the achievement of the European energy and climate targets.

Energy transition to zero carbon emission cities requires also energy use in a smart way, so as to reduce the energy degradation in terms of capacity to generate useful work. Thus, in order to improve the energy efficiency, it is essential to find low carbon solutions for the complex energy systems of urban environments (i.e. industrial, transport sector) that will increase cities' overall energy and resource efficiency. Taken into consideration the fact that available standards do not provide integrated methods for the assessment of smart cities [ISO, 2014], the development of a comprehensive and holistic approach that will address policies for sustainable development and planning of smart cities is of great importance for climate change mitigation. It is noted, that in order to evaluate the energy and resource flows of different urban energy systems, it is necessary that these be analyzed not only within the city but also across city boundaries. Changes over time as a result of socioeconomic and technical strategies need to be examined and monitored, so as to quantify the progress towards the creation of low carbon cities and to visualize results in ways that are usable by stakeholders.

In this direction, exergy analysis is a useful tool for the evaluation of the effectiveness of different energy systems in smart cities. As per second law of thermodynamics, different forms of energy have different energy potential to generate useful work; therefore, exergy is used in order to measure the degradation of energy in conversion processes. Exergy analysis can be used, in order to identify how smartly energy is used. Until now exergy analysis has been applied in comparative analysis of countries, regions, economic, industrial as well as transportation sectors [Dincer and Rosen, 2013; Koroneos and Nanaki, 2008]. Nonetheless, the smartness of energy policies aiming to optimize their efficiency has not yet been investigated. Furthermore, many energy policies related to decision-making fail to take into consideration life-cycle impacts (both economic and environmental) of energy efficiency measures in the urban environment. Therefore, a holistic approach is necessary, in order to tackle climate change mitigation.

In order to bridge the above mentioned gaps, this book addresses, through a holistic approach - future energy needs of smart cities. The transportation as well as the energy system of the future has to be intelligent, sustainable and completely integrated in the multi-modal logistics chain of the future.

For this reason, the methods of exergy analysis as well as of life cycle analysis are being employed, in order to support smart city planning, in terms of evaluating the exergy efficiency of renewable energy systems that are going to be integrated to the electricity market and managed on the exploitation of resources within smart cities. In order to improve the energy efficiency of complex energy systems (i.e. transportation, built environment), it is imperative to find low carbon solutions. For this reason, this book also throws light on the correlation of the exergy concept in relation to the energy and transportation sector aiming to increase the efficiency of this sector in relation to its surroundings, within the concept of this city. In this direction, exergy analysis is combined with life cycle analysis.

The basics of exergy analysis as well as of sustainable smart cities are presented; however the fundamental concept of exergy applied to the above mentioned sectors is the basic axis of this book (Fig. 1.1). Transportation in urban environment and interaction with the intelligent energy systems is studied. Exergy analysis is used within regional planning; referring to sustainable spatial design (improved use of energy flows, energy demand, e.g.). Moreover, possibilities through intelligent energy systems implemented are presented for energy conversion, distribution and storage to increase the integration of sustainable energy to the system and deliver energy to the end user under the most suitable scheme. An exergy planning approach focuses towards that direction and examines smart grids, electricity markets and how much they interact with each other. Big Data, Deep Learning, and Nexus Platforms are recruited to demonstrate the amount of data being wasted from a number of processes at an urban scale and collect them, organize and utilize them under a "smart city v2.0" version. In addition, the term of Bi-directional exergetic efficiency (BiXef) is introduced so as to measure not only how efficient a resource is, but mainly if it is as efficient when needed.

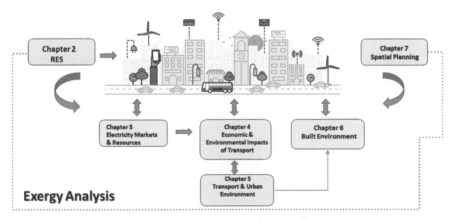

Fig. 1.1. Flow chart of the book

Color version at the end of the book

1.2 Motivation for the Book

Intelligent Energy Systems within the scope of smart cities are gaining recognition as alternative ways through which a low carbon society can be pursued. This term is drawing increasing attention from urban planners and decision-makers as it emphasizes the need for enhancing cities' capacities to cope with the heterogenous challenges of cities and their future development and climate change. Therefore, a holistic approach to urban development is of great significance, especially in the case of contemporary cities, and what this entails in terms of sustainability and the integration of its environmental, social, and economic dimensions. Theoretical knowledge, and available research tools such as exergy analysis, should contribute to make urban development smartly more sustainable. Based on the above mentioned, the motivation for this book lies in the following areas:

- The significance of the interdisciplinary field of smart cities as it requires a multi-disciplinary approach starting from conceptual designs (modeling) to the implementation of this within the city.
- The determination of what has been done until now in the area of optimization of complex energy systems within smart cities.
- The presentation of an overview of key concepts, theories, and discourses in regards to exergy analysis and energy systems within smart cities.
- The assessment and synthesis of the existing information in line with the concept of exergy analysis.
- To provide a solid background and theoretical foundation for exergy analysis and renewable energy systems within smart cities.
- To identify the gaps in the existing research that this book is endeavoring to address, positioning this book in the context of previous research and creating a research space for it.
- To produce a rationale and establish the need for this study and thus justify its originality.
- To bridge the gap between academic understanding of exergy analysis and ecosystem functions and municipal planning of organizational structures.

1.3 Aim and Objectives of the Book

The concept of sustainable energy planning is interconnected via the method of exergy analysis with the smart cities' perspective. In this direction, this book constitutes a valuable tool that can be helpful to stakeholders and decision-makers in the process of regional planning, referring to sustainable spatial design. Thus, it aims to help in decision making, especially with regards to national and sustainable energy designs, and demonstrates how intelligent energy systems can best be utilized. The book's objective is to provide useful insight to policy-makers, urban planners, engineering consultancies, scientists, researchers, students as well as citizens interested for optimization of energy systems within intelligent energy urban areas.

This book addresses the concept of exergy analysis within sustainable energy planning and contributes to the literature regarding optimization of energy systems within smart cities. Methodologies and theoretical foundations of exergy analysis as well as case studies within the concept of smart cities are presented. Furthermore, the book aims to point out that a combination of efficiency measures, intelligent load management and local renewables, can contribute to a decarbonized urban energy future. In the book, several scales within the city are addressed, ranging from the efficiency of renewable energy systems to complex energy systems such as the transportation sector.

1.4 Structure of the Book

The book discusses the exergetic aspects of renewable energy system in the key areas of transport and energy sector, within the concept of smart cities. The book is structured into six sections that are briefly described here below.

In **Chapter 2**, the background and the method of exergy analysis is presented. It establishes the fundamental notion of energy – exergy and thermodynamics with application on Renewable Energy Systems. The chapter provides the necessary background for understanding these concepts, as well as basic principles, general definitions and practical applications and implications. Case studies (solar energy, wind energy and hybrid energy systems) are provided to point out aspects of energy, entropy and exergy. The scope of this chapter is to emphasize on the role of exergy analysis as a tool for the creation of a sustainable energy future by increasing the energy efficiencies of processes utilizing sustainable energy resources.

Chapter 3 points out the need for the improvement of the performance of smart cities via a holistic system operation, which shall lead to increased efficiencies of the system and reduced operational costs. By improving the efficiency in urban areas via a holistic approach, will lead to increased performances of all urban related activities squeezing as much as possible the operational costs, and goes a lot further than just the optimization of electricity markets. However, optimizing electricity markets is the starting point for city planners and decision-makers towards achieving cleaner investments and monitor at the end a unified smart city operation.

Chapter 4 then deals with available Alternative Fuel Vehicles and the renewable energy sources that can be employed for the creation of a low carbon transport sector within smart cities. The chapter focuses on the key challenges for smart mobility consideration in smart city – in terms of economic and environmental assessment. The main questions are which transport systems will be used in the future – in terms of cost and environmental efficiency. It is argued that alternative vehicle technologies within smart cities have to be considered as part of an integrated energy system. Based on real data from the vehicle industry, the total cost of ownership as well as the life cycle environmental impacts of different powertrains are assessed.

In **Chapter 5**, transport energy demand is discussed. The interrelation between urban environment and the transportation sector as well as between climate change and the need for sustainable transport systems is presented. Optimization of the transport sector is proposed via the method of exergy analysis. The optimization is then applied to Greece's transportation sector. Using detailed data for the period 1980-2016, the efficiency of the transport system is assessed. The assessment includes a detailed transport system analysis - in terms of energy and exergy efficiency - and discusses how the system (road, rail, aviation and navigation) can be optimized with the integration of alternative fuels (i.e. electricity and biofuels).

Chapter 6 focuses on the Urban Nexus Platform which should serve as a tool for organizing small scale resources within a city level by measuring efficiencies within one energy system. The platform will optimize the true potentials of new energy technologies using artificial intelligence and deep learning. An index, measuring when a resource within a city/neighborhood is needed or not, is introduced. In addition, this index quantifies how much of the resource is needed. The Bi-directional exergetic efficiency (BiXef) is introduced to measure not only how efficient a resource is, but mainly if it is as efficient when needed. This approach, when adopted, will change the way we evaluate global resources.

In **Chapter 7**, the uncertainty of spatial planning within the concept of smart cities is assessed using an exergetic approach. The chapter discusses design measures for the optimization of spatial planning and how GIS systems could improve systems' efficiencies. Spatial planning and remote sensing and their large number of applications are among the most known tools that all planners currently use to identify spots where system losses are (sometimes) falsely neglected within cities. The presentation of the transportation-energy-food nexus presented there shows that there is room for improvement in the way we work now on integrated urban systems.

In summary, the book analyzes the key challenges of energy transition to zero carbon cities and the methods to provide solutions – in terms of system's optimization- are discussed. It includes an analysis of the existing transport and energy efficiency options and the potentials of renewable energy systems within the concept of intelligent communities. Exergy analysis is used to evaluate energy performance of energy and transport systems. As a result, the transition to a zero-carbon future is outlined, and innovative tools to model, monitor, and optimize performance are presented. The main difference as well as the main advantage of this book, compared to other books, lies in the fact that for the first time, the concept of sustainable energy planning is interconnected via the method of exergy analysis with the smart cities' perspective. In this direction, this book constitutes a valuable tool that can be helpful to stakeholders and decision-makers in the process of regional planning; referring to sustainable spatial design.

REFERENCES

Dincer, I. and Rosen, M. (2013). Exergy Analysis of Countries, Regions, and Economic Sectors. DOI: 10.1016/B978-0-08-097089-9.00021-8. *In*: Exergy, pp. 425–450.

Doelle, M. (2016). The Paris agreement: historic breakthrough or high stakes experiment? Clim. Law, 6 (1–2): 1-20.

International Organization for Standards – ISO (2014). Sustainable development of communities – Indicators for city services and quality of life. ISO 37120. Geneva.

International Energy Agency – IEA: International Energy Outlook (2017). Available online: https://www.eia.gov/outlooks/ieo/pdf/0484 (2017).pdf

Koroneos, C. and Nanaki, E. (2008). Energy and exergy utilization assessment of the Greek transport sector. Journal of Resources Conservation and Recycling 52(5): 700–706.

Basics of Renewable Energy Systems Exergetics

E. Nanaki[1] and G. Xydis[2]

[1] University of Western Macedonia, Department of Mechanical Engineering, Bakola & Sialvera, Kozani 50100, Greece
[2] Department of Business Development and Technology, Aarhus University, Birk Centerpark 15, 7400 Herning, Denmark

2.1 Introduction

Since the industrial revolution, coal has been considered as one of the main energy sources. Much of today's global industrial production is based on energy from burning coal. It is also generally considered as a cheap fuel with stable prices. However, the burning of coal produces carbon dioxide emissions that affect global warming, despite the fact that climate change continues to have numerous opponents. In regards to oil, it is noted that the global oil production has been intensified since the mid-19th century and from the mid-20th century the production rated increased rapidly. Today, after two oil crises (1973 and 1979) the impacts of oil exploitation are even more evident. Oil can be characterized by its high-energy density and its ability for easy production and refining. It is a stable power source and the technology of producing electricity from oil is considered mature. Nonetheless similar to coal, oil produces Greenhouse Gas Emissions (GHG) when burned. Among the disadvantages we can add the environmental impact that may occur by oil spills and potential terrorist acts. To be more specific, the utilization of energy resources such as fossil fuels emits significant amounts of pollutants causing serious environmental problems such as global warming, ozone layer depletion as well as climate change.

The creation of a resilient Energy Union with a climate change policy that is capable to achieve the adopted 2020 and 2030 climate targets and secure energy supply is among the European Union's long-term climate and energy objectives (EU). In order to achieve this, Europe has to decarbonize its energy supply, integrate the fragmented national energy markets into a smooth functioning and coherent European system, and set up a framework that allows the effective coordination of national efforts [EC, 2015]. Hence, the need for a sustainable development agenda that promotes the use of cleaner forms of energy that can promote human wellbeing, economic development and environmental protection is of great significance. Renewable Energy Sources (RES) can play a vital role to this energy transition as they can mitigate Greenhouse Gas Emissions (GHG) and can contribute to the reduction of the reliance of fossil fuels. Other benefits associated with the growth of RES include the reduction of fossil fuel imports, the diversification of energy supply and the creation of jobs, skills and innovation in local markets and in progressive sectors with significant growth potential. In a life-cycle perspective, the environmental pressures arising from renewable energy technologies are 3–10 times lower than from fossil fuel based systems [UNEP, 2015].

Renewable energy is derived from natural processes that can be replenished within a short time scale and can be derived directly or indirectly from the sun and from other natural mechanisms [www.treia. org]. Renewable energy sources include hydropower, bioenergy, thermal, geothermal, wind, photochemical, photoelectric, tidal, wave, and solar energy. It excludes energy from fossil fuel sources (oil, coal and natural gas). Renewable energy does not include energy resources derived from fossil fuels. Renewable energy does not include energy resources derived from fossil fuels, waste products from fossil sources, or waste products from inorganic sources.

Renewable energy technologies are considered to be less competitive than traditional electric energy conversion systems, mainly because of their intermittency and the relatively high maintenance cost. However, renewable energy sources have several advantages, such as the reduction in dependence on fossil fuel resources and the reduction in carbon emissions to the atmosphere. Furthermore, renewable energies avoid the safety problems derived from atomic power [Strupczewskim, 2003], which is why, from a social point of view, it has become more desirable to adopt renewable energy power plants [Skoglund et al., 2010]. Nonetheless as the renewable energy systems suffer from low conversion efficiency the use of them requires special consideration. Taking into account that exergy analysis provides a useful tool in assessing and comparing processes and systems, it can be deducted that exergy analysis can be employed, in order to improve and optimize renewable energy systems.

Exergy analysis is a thermodynamic analysis technique based primarily on the Second Law of Thermodynamics. As an alternative to energy analysis, exergy analysis constitutes a tool, which can be used, in order to assess and compare processes and systems. Exergy can be considered to be a measure of the maximum useful work that can be done by a system interacting with an environment which is at a constant pressure P_0 and a temperature T_0. Consequently, exergy analysis can assist in improving and optimizing designs. Two key features of exergy analysis are [Brodyanski et al., 1994] it yields efficiencies which provide a true measure of how nearly actual performance approaches the ideal, and [Kotas, 1995] it identifies more clearly than energy analysis the types, causes and locations of thermodynamic losses.

Exergy is defined as the amount of work available from an energy source. Exergy is the useful energy that can be exploited from an energy resource or a material, which is subjected in an approximately reversible procedure, from an initial situation till balance with the natural environment is restored. Exergy is dependent on the relative situation of a system and its ambient conditions, as they are determinate by a sum of parameters, and it can be equal with the zero (in a balanced situation with the environment) [Szargut et al., 1988]. Exergy is a measurement of how far a certain system deviates from a state of equilibrium with its environment. The exergy concept incorporates both the quantitative and qualitative properties of energy. Every irreversible phenomenon causes exergy losses leading to exergy destruction of the process or to an increased consumption of energy from whatever source the energy was derived.

Based on the above mentioned exergy analysis can be considered as a method, used for the evaluation of maximum work extractable from a substance relative to a reference state (i.e., dead state). This reference state is arbitrary, but for terrestrial energy conversion the concept of exergy is most effective if it is chosen to reflect the environment on the surface of the Earth. The various forms of exergy are due to random thermal motion, kinetic energy, potential energy associated with a restoring force, or the concentration of species relative to a reference state. In order to establish how much work potential a resource contains; it is necessary to compare it against a state defined to have zero work potential. An equilibrium environment which cannot undergo an energy conversion process to produce work is the technically correct candidate for a reference state [Herman, 2006].

Increasing application and recognition of the usefulness of exergy analysis in industry, government and academia has been observed during the last decades. In this direction, applications of exergy analysis have occurred in a diverse range of fields, including electricity generation and cogeneration, fuel processing, energy storage, transportation, industrial energy use, building energy systems, and others. In addition, exergy analysis has been applied to fields outside thermodynamics, such as biology and ecology, management of industrial systems and economics. To be more specific, exergy analysis has been employed in the design, simulation and performance evaluation of energy systems, industrial systems [Rosen and Dincer, 2005; Xydis et al., 2011; Rosen and Dincer, 2004; Koroneos et al., 2011], thermal energy storage [Rosen et al., 2004], countries [Wall, 1990; Wall et al., 1994; Chen and Chen, 2009; Rosen and Dincer, 1997], transportation systems [Koroneos and Nanaki, 2011; Jaber et al., 2008; Gasparatos et al., 2009] as well as of environmental impact assessments [Crane et al., 1992; Gunnewiek and Rosen, 1998].

In this chapter, the employment of exergy analysis to renewable energy systems is presented and analyzed. Theoretical aspects of energy and exergy analysis are described, including fundamental principles as well as general relations. Then two case studies involving the application of exergy analysis to renewable energy systems are investigated: solar energy systems and wind energy systems.

2.2 Energy and Exergy Modeling

The objective of exergy analysis is to determine the exergy losses (thermodynamic imperfections) and to evaluate quantitatively the causes of the thermodynamic imperfection of the process under consideration. Exergy analysis can lead to all kinds of thermodynamic improvement of the process under consideration [Szargut et al., 1988]. It is noted that exergy's analysis main advantage is that it connects the real output with the theoretical (ideal) one. Even if the theoretical maximum cannot be reached, it provides a point of comparison for the further possibilities of a procedure optimization. For a general steady state, steady flow process system the mass balance equation can be expressed by Eq. (1)

$$\sum m_{in} = \sum m_{out} \tag{1}$$

where m is the mass flow rate, and the subscript *in* stands for inlet and *out* for outlet. The general energy balance is expressed by Eq. (2):

$$\sum E_{in} = \sum E_{out} \tag{2}$$

where E_{in} is the rate of net energy transfer in, E_{out} is the rate of net energy transfer out by heat, work and mass [Caliskan, 2015].

In the absence of electricity, magnetism, surface tension and nuclear reaction, the total exergy of a system $\dot{E}x$ is divided into four components, namely (i) physical exergy $\dot{E}x^{PH}$, (ii) kinetic exergy $\dot{E}x^{KN}$, (iii) potential exergy $\dot{E}x^{PT}$ and (iv) chemical exergy $\dot{E}x^{CH}$.

$$\dot{E}x = \dot{E}x^{PH} + \dot{E}x^{KN} + \dot{E}x^{PT} + \dot{E}x^{CH} \tag{3}$$

Inspite of the fact that exergy is an extensive property, it is convenient to work with it on a unit of mass or molar basis. The total specific exergy on a mass basis is written as follows:

$$Ex = Ex^{PH} + Ex^{KN} + Ex^{PT} + Ex^{CH} \tag{4}$$

The general exergy balance can be written as

$$\sum \dot{E}x_{in} - \sum \dot{E}x_{out} = \sum \dot{E}x_{dest} \tag{5a}$$

or

$$\dot{E}x_{heat} - \dot{E}x_{work} + \dot{E}x_{mass,in} - \dot{E}x_{mass,out} = \dot{E}x_{dest} \tag{5b}$$

with

$$\dot{E}x_{heat} = \sum (1 - \frac{T_o}{T_k})Qk \tag{6a}$$

$$\dot{E}x_{work} = W \tag{6b}$$

$$\dot{E}x_{mass,in} = \Sigma_{min\psi out} \tag{6c}$$

$$\dot{E}x_{mass,out} = \Sigma_{mout\psi out} \tag{6d}$$

where Qk is the heat transfer rate through the boundary at temperature T_k at location k and W is the work rate. The specific exergy is calculated as:

$$ex_{tm} = (h - h_0) - T_0(S - S_0) \tag{7}$$

where h is enthalpy, S is entropy, and the subscript zero indicates properties at the restricted dead state of P_0 and T_0. The rate form of the entropy balance can be expressed as:

$$S_{in} - S_{out} + S_{gen} = 0 \tag{8}$$

where the rates of entropy transfer by heat transferred at a rate of Qk and mass flowing at a rate of m are $S_{heat} = Q_k/T_k$ and $S_{mass} = ms$, respectively. Taking the positive direction of heat transfer to be to the system, the rate form of the general entropy relation given in Eq. (8) can be rearranged to give:

$$S_{gen} = \sum m_{out}S_{out} - \sum m_{in}S_{in} - \sum \frac{Qk}{T_k} \tag{9}$$

It is usually more convenient to find S_{gen} first and then to evaluate the exergy destroyed or the irreversibility rate I directly from Eq. (10) [Szargut,

2005]. For a process occurring in a system, the difference between the total exergy flows into and out of the system, less the exergy accumulation in the system, is the exergy consumption I, expressible as

$$I = \dot{E}x_{\text{dest}} = T_o S_{\text{gen}} \tag{10}$$

The **specific exergy** of an incompressible substance (i.e., water) is calculated as follows [Szargut, 2005]:

$$\psi_w = C \left(T - T_0 - T_0 \ln T/T_0 \right) \tag{11}$$

The total flow exergy of air is calculated as below [Wepfer et al., 1979] $\Psi_{a,t} = (C_{p,a} + w C_{p,v}) T_o [(T/T_o) - 1 - \ln (T/T_o)] + (1+1.6078w) Ra\, T_o \ln (P/P_o) +$

$$Ra T_o \left\{ \frac{(1 + 1.607w)\ln(1 + 1.6078wo)}{(1 + 1.6078w) + \left(1.6078w \ln\left(\dfrac{w}{wo} \right)\right)} \right\} \tag{12}$$

where the specific humidity ratio is

$$w = m_v / m_a \tag{13}$$

Assuming air to be a perfect gas, the specific physical exergy of air is calculated as below [Kotas, 1995]:

$$\Psi_{a,\,per} = C_{p,\,a} \left(T - T_o - T_o \ln T/T_o \right) + Ra T_o \ln P/P_o \tag{14}$$

2.2.1 Reference Environment

Exergy is always evaluated with respect to a reference environment (i.e. dead state). For this reason, the properties of the reference environment determine the exergy of a stream or a system. When a system is in equilibrium with the environment, the state of the system is called the dead state due to the fact that the exergy is zero. At the dead state, the conditions of mechanical, thermal, and chemical equilibrium between the system and the environment are satisfied: the pressure, temperature, and chemical potentials of the system equal those of the environment, respectively. In addition, the system has no motion or elevation relative to coordinates in the environment. Under these conditions, there is neither the possibility of a spontaneous change within the system or the environment nor an interaction between them.

Taking into consideration the fact that chemical reactions cannot occur between the environmental components, it can be considered that the reference environment acts as an infinite system, and constitutes a source for heat and materials. It experiences only internally reversible processes in which its intensive state remains unaltered (i.e., its temperature T_o, pressure P_o and the chemical potentials μ_{io} for each of the i components present remain constant). The exergy of the reference environment is zero. The exergy of a stream or system is zero when it is in equilibrium with the reference environment. Taking into account the fact that the natural environment is not in equilibrium, and its intensive properties present spatial and temporal variations, it cannot be considered as a reference environment. Many chemical reactions in the natural environment are blocked because the transport mechanisms necessary to reach equilibrium are too slow at ambient conditions. From the above mentioned it is deducted that the exergy of the natural environment is not zero and work could be obtained if it were to come to equilibrium. Different models regarding the reference environment are employed, in order to achieve a balance between the theoretical requirements of the reference environment and the actual behavior of the natural environment. These can be summarized as follows:

- **Natural-environment-subsystem models:** These models try to simulate realistically subsystems of the natural environment and represent a significant number of reference environment models. These include inter alias the models suggested by Baehr and Schmidt [1963] as well as these proposed by Gaggioli and Petit [1977] and Rodriguez [1980]. The temperature and pressure of this reference environment are considered to be 25°C and 1 atm, respectively, and the chemical composition is assumed to consist of saturated air with vapor water, and the following condensed phases at 25°C and 1 atm: water (H_2O), gypsum ($CaSO_4 \cdot 2H_2O$) and limestone ($CaCO_3$).

- **Reference-substance models:** These models are based on the selection of a "reference substance" and the assignment of zero exergy for every chemical element. One such model is the one proposed by Szargut [1967]. In this model, the reference substances are selected as the most valueless – found in abundance in the natural environment. Part of this environment is the composition of moist air, including N_2, O_2, CO_2, H_2O and the noble gases; gypsum (for sulfur) and limestone (for calcium).

- **Equilibrium models:** In this type of model all the materials present in the atmosphere, oceans and a layer of the crust of the Earth are pooled together and an equilibrium composition is calculated for a given temperature [Ahrendts, 1980]. The selection of the thickness of crust considered is subjective and is intended to include all materials accessible to technical processes. Ahrendts considered thicknesses varying from 1 m to 1000 m, and a temperature of 25°C.

- **Constrained-equilibrium models:** A modified version of equilibrium model suggests the calculation of an equilibrium composition excluding the possibility of the formation of nitric acid (HNO_3) and its compounds [Ahrendts, 1980]. That is, all chemical reactions in which these substances are formed are in constrained equilibrium, and all other reactions are in unconstrained equilibrium. When a thickness of crust of 1 m and temperature of 25°C were used, the model was similar to the natural environment.

- **Process-dependent models:** This model takes into consideration components in a stable equilibrium composition at the temperature and total pressure of the natural environment [Bosnjakovic, 1963]. This model is dependent on the process examined, and is not general. Exergies evaluated for a specific process-dependent model are relevant only to the process; they cannot rationally be compared with exergies evaluated for other process-dependent models.

2.2.2 Energy and Exergy Efficiency

Efficiency is defined as "the ability to produce a desired effect without waste of, or with minimum use of, energy, time, resources, etc." In regards to resource utilization, efficiency is of great significance. For general engineering systems, non-dimensional ratios of quantities are typically used to determine efficiencies. Ratios of energy are conventionally used to determine efficiencies of engineering systems whose primary purpose is the transformation of energy. These efficiencies are based on the First Law of Thermodynamics. A process has maximum efficiency according to the First Law if energy input equals recoverable energy output (*i.e.*, if no "energy losses" occur). However, efficiencies determined using energy is misleading because in general they are not measures of "an approach to an ideal". Energy efficiency of steady-state processes occurring in systems is described in Eq. (15):

$$n_i = \frac{\sum E_{out}}{\sum E_{in}} \tag{15}$$

where "out" stands for "net output" or "product" or "desired value", and "in" stands for "given" or "used" or "fuel".

Taking into consideration the fact that exergy is a measure of the ability to perform work, maximum efficiency is attained for a process in which exergy is conserved. Therefore exergy efficiency (Second Law of Thermodynamics) is of great significance in energy utilization. Despite the fact that there is no standard set of definitions in the literature, two different approaches are generally used ("brute-force" and "functional") [DiPippo, 2004]. A "brute-force" exergy efficiency for any system is defined as the ratio of the sum of all output exergy terms to the sum of all input exergy terms. A "functional" exergy efficiency for any system is defined as the ratio of the exergy associated with the desired energy output to the exergy associated with the energy expended to achieve the desired output. Exergy efficiencies are often more intuitively rational than energy efficiencies because efficiencies between 0 and 100% are always obtained. Measures which can be greater than 100% when energy is considered, such as coefficient of performance, normally are between 0 and 100% when exergy is considered. In fact, some researchers [Gaggioli, 1983] call exergy efficiencies "real" or "true" efficiencies, while calling energy efficiencies "approximations to real" efficiencies.

Different ways of formulating exergetic efficiency proposed in the literature have been given in detail elsewhere [Cornelissen, 1997]. The exergetic efficiency expresses all exergy input as used exergy, and all exergy output as useful exergy. Therefore, the exergetic efficiency is expressed by Eq. (16):

$$n_{ii} = n_{ex} = \frac{Ex_{out}}{Ex_{in}} \tag{16}$$

Exergy efficiency is improved to the maximum when the exergy loss or irreversibility $Ex_{in} - Ex_{out}$ is minimized [Van Cool, 1997]. When heat at a temperature $T_t > T_o$ is available and the temperature of a thermodynamic system is T_t and the temperature of the environment is T_o its exergy quality can be expressed by the quality factor (Eq. 17):

$$\text{Quality factor} = 1 - \frac{T_o}{T_t} \tag{17}$$

When the temperature T_t, at which the heat is available, increases, the quality factor increases too. The **irreversibility** or the **exergy loss** of an open system in steady state is given by:

$$I = \sum_i (1 - \frac{T_o}{T_i})Q_i - W_{net} + \sum_{in}(me)_{in} - \sum_{out}(me)_{out} \tag{18}$$

where T is the temperature of environment, Q_i is the heat transportation rate through a system's boundaries for a constant temperature T_i, W_{net} is the rate of work production, m is the mass flow rate of every flow of materials that enters the system and e is the specific exergy of the flows. The **specific exergy** e constitutes of thermo mechanical, e_{tm} and chemical e_{ch} exergy.

Equation (18) indicates that during a thermodynamic process, a system goes from an initial state to a final state and one or more of its thermodynamic properties change. Ideally, a process can be reversed completely and the system can be restored to its initial state without a trace that shows that it went through a thermodynamic change. To have a reversible process, all the steps from the initial to final state should be reversible. However, real processes are not reversible and occur due a finite gradient that subsists between two states of the system (e.g. heat transfer between two bodies of a different temperature). The internal exergy loss is attributed to irreversibilities that accompanied real processes and it can be calculated as the difference between the sum of the exergy value of the ingoing flows and the sum of the exergy value of the outgoing flows. The larger the gradient between the two states, the larger the exergy loss.

Exergy analysis is employed to detect and to evaluate quantitatively the causes of thermodynamic imperfection of the process under consideration and is indicative of the possibilities of thermodynamic improvement of the process under consideration [Hammond and Stappleton, 2001; Koroneos et al., 2003; Becali et al., 2003].

2.2.3   Material Properties

In order to perform energy and exergy analysis it is imperative to have data regarding the material properties. Property data are widely available for many substances (e.g., steam, air and combustion gases as well as chemical substances). Energy values of heat and work flows are absolute, while the energy values of material flows are relative. Enthalpies are evaluated relative to a reference level. Taken into consideration the fact that energy analyses deal with energy differences, the reference level used for enthalpy calculations can be arbitrary. For the determination of some energy efficiencies, however, the enthalpies must be evaluated relative to specific reference levels, e.g., for energy-conversion processes, the reference level is often selected so that the enthalpy of a material equals its Higher Heating Value (HHV).

Nonetheless in the case of a comparative energy and exergy analysis, it is imperative to clarify reference levels for enthalpy calculations. In this way, the enthalpy of a compound is going to be evaluated in relation to the stable components of the reference environment. Thus, a compound that exists as a stable component of the reference environment is considered to have an enthalpy of zero at T_0 and P_0. Enthalpies calculated with respect to such conditions are mentioned as "**base enthalpies**" [Rodriguez, 1980]. The base enthalpy is similar to the enthalpy of formation. While the latter is the enthalpy of a compound (at T_0 and P_0) relative to the elements (at T_0 and P_0) from which it would be formed, the former is the enthalpy of a component (at T_0 and P_0) relative to the stable components of the environment (at T_0 and P_0). For many environment models, the base enthalpies of material fuels are equal to their HHVs. Table 2.1 summarizes base enthalpies for many substances.

In order to perform exergy analysis, it is imperative to calculate relevant chemical exergy values. In this direction, different methods have been developed for the evaluation of chemical exergies of solids, liquids and gases [Rodriguez, 1980; Sussman, 1980]. It is noted that for

Table 2.1. Base enthalpy and chemical exergy values of selected species
[Gaggioli and Petit, 1997; Rodriguez, 1980]

Substances	Chemical formula	Specific base enthalpy (kJ/g-mol)	Specific chemical exergy* (kJ/g-mol)
Ammonia	NH_3	382.585	$2.478907 \ln y + 337.861$
Carbon – graphite	C	393.505	410.535
Carbon dioxide	CO_2	0.000	$2.478907 \ln y + 20.108$
Carbon monoxide	CO	282.964	$2.478907 \ln y + 275.224$
Ethane	C_2H_6	1,564.080	$2.478907 \ln y + 1,484.952$
Hydrogen	H_2	285.851	$2.478907 \ln y + 235.153$
Methane	CH_4	890.359	$2.478907 \ln y + 830.212$
Nitrogen	N_2	0.000	$2.478907 \ln y + 0.693$
Oxygen	O_2	0.000	$2.478907 \ln y + 3.948$
Sulfur	S	636.052	608.967
Sulfur dioxide	SO_2	339.155	$2.478907 \ln y + 295.736$
Water	H_2O	44.001	$2.478907 \ln y + 8.595$

y stands for the molal fraction for each of the respective species.

complex materials, such as coal, ash etc, approximation methods have been developed. By considering environmental air and gaseous process streams as ideal gas mixtures, chemical exergy can be calculated for gaseous streams using component chemical exergy values.

2.3 Exergetic Analysis and Evaluation of Renewable Energy Systems

Energy and exergy analyses can be performed for different energy systems – including Renewable Energy Systems. The procedure that should be followed includes the following steps:

- Breakdown of the process under consideration into as many sections as necessary.
- Carrying out conventional mass and energy balances of the process. Calculation of all basic quantities (e.g., work, heat) and properties (e.g., temperature, pressure).
- Selection of a reference–environment model. The selection of the reference model should take into consideration the nature of the process as well as the acceptable degree of analysis complexity and accuracy.

- Evaluation of the energy and exergy values, relative to the selected reference-environment model.
- Calculation of exergy consumptions.
- Selection of efficiency definitions and evaluation values for the efficiencies.
- Interpretation of results. Conclusions and recommendations, relative to optimizing system design.

Based on the above mentioned procedure and on the method of exergy and exergy analysis – as this was presented in previous sections–, case studies of evaluating different Renewable Energy Systems are going to be presented. This section focuses on renewable energy systems and their components and discusses the use of exergy analysis to assess and improve them. Exergy methods provide a physical basis for understanding, refining and predicting the variations in renewable behavior.

2.3.1 Solar Energy Systems

Solar energy is an available, cheap and environmentally friendly alternative source of energy, which can be integrated with different kinds of systems, in order to reduce energy consumption. Solar energy can be exploited via solar thermal technologies, which be used for water heating, space heating, space cooling, water treatment and process heat generation. In addition, solar energy can be used for the production of electric energy. Solar power converts sunlight into electricity, either directly using photovoltaics (PV), or indirectly using Concentrated Solar Power (CSP). CSP systems use lenses or mirrors and tracking systems to focus a large area of sunlight into a small beam. PV converts light into electric current using the photoelectric effect.

During the last decades the renewable and in some cases the economic features of solar energy systems have attracted attention from political and business decision makers and individuals. The above, in conjunction with the increased usage of PV applications in many countries and in different sectors (residential, commercial and industrial), point out the necessity of investigating the exergy efficiency of such systems. Figure 2.1 indicates that there are different ways to produce electricity from the solar energy. Solar radiation is used mainly in three different ways as Thermal Production (TP), Electric Production (EP) and Electric and Thermal Production.

It should be noted that the evaluation of the performance of solar energy systems (with the use of exergy analysis method) is a difficult procedure, requiring the calculation of the exergy of radiation. The difficulty lies in the fact that exergy stands for the maximum quantity of work that can be produced in some given environment (usually the terrestrial environment, considered as an infinite heat source or sink).

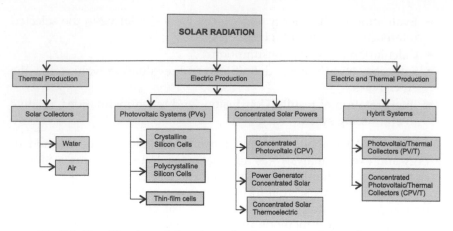

Fig. 2.1. Classification of electric production systems from solar energy

The proposed maximum efficiency ratio (or exergy-to-energy ratio for radiation) for determining an exergy of thermal emission at temperature T, is as follows [Petela, 2003; Candau, 2003]:

$$\Psi_{srad,max} = 1 + \frac{1}{3}(T_0/T)^4 - \frac{4}{3}(T_0/T) \tag{19}$$

where T is considered to be equal to the solar radiation temperature (T_s) with 6000K in exergetic evaluation of a solar cylindrical-parabolic cooker and a solar parabolic-cooker by Petela [2005] and Ozturk [2004], respectively. This result has been also obtained independently by Landsberg and Mallinson [1976]. It has been shown that the interaction between two black bodies at different temperatures involves irreversibilities in both black bodies [Beretta and Gyftopoulos, 2015]. The exergy factor of the radiation emitted by a source of geometric factor f has been calculated as [Badescu, 2014]:

$$\psi = 1 - \frac{4}{3}\frac{T_0}{T} + \frac{1}{3}\frac{1}{f}\left(\frac{T_0}{T}\right)^4 \quad f \geq \left(\frac{T_0}{T}\right)^3 \tag{20}$$

When $f < (T_0/T)^3$ work cannot be extracted from radiation energy.

As already mentioned in the previous sections, exergy analysis not only can be used, so as to evaluate the efficient usage of solar energy systems, but also to improve their efficiency (by determining the sources and magnitude of irreversibilities). In this direction, a number of studies have been conducted in evaluating different solar energy systems. These studies refer to flat-plate solar collectors [Ge et al., 2014; Fudholi et al., 2013; Oztop et al., 2013]; combined photovoltaic and thermal (PV/T) collectors [Zhang et al., 2014; Dupeyrat et al., 2014; Kasaeian et al., 2013]; parabolic trough collectors [Hou et al., 2014; Padilla et al., 2014]; as well as parabolic dish collectors [Madadi et al., 2014].

Furthermore, the energy content of the solar radiation can be utilized in different applications and processes. The solar radiation, which is converted into thermal energy, can be used for various purposes, such as the direct generation of space heating, hot water or even process heat (i.e. desalination). As the analysis and optimization of such applications is of great interest, many studies have taken place. These studies analyze solar collectors coupled with Phase Change Materials (PCMs) that store solar thermal energy in the form of latent heat [Yang et al., 2014]. Also in other cases, the solar heat, which is usually delivered to a heat transfer fluid, can be also provided to absorption cooling systems or heat pumps, for the indirect production of cooling, especially for buildings. In this direction, many studies have performed exergy analysis for solar drying system [Fudholi et al., 2014]; hybrid Li/Br absorption cooling system [Gomri, 2013]; and solar-driven desalination system based on humidification-dehumidification technology [Nematollahi et al., 2013]. Finally, several applications aim at the production of electricity as the final product through the implementation of power cycles (i.e. Organic Rankine Cycle and the Kalina Cycle). The production of electricity is sometimes achieved directly, through photovoltaic systems. The generated electricity can be sold to the power grid or it can be used either for the operation of heat pumps for other purposes, such as the electrolysis of water to produce hydrogen. Exergy analysis of such systems has also been performed in the studies of Wang et al. [2014]; Rovira et al. [2013] as well as of Sun et al. [2014].

2.3.1.1 *Energy and Exergy Analysis of Solar Photovoltaic Systems*

Solar photovoltaic (PV) technology converts sunlight directly into electrical energy. Direct current electricity is produced, converted to alternating current or stored for later use. Solar PV systems operate in an environmentally benign manner, have no moving components, and have no parts that wear out – provided that the device is protected from the environment. A PV cell is a type of photochemical energy conversion device. Others include photoelectric devices and biological photosynthesis. Such systems operate by collecting a fraction of the radiation within some range of wavelengths. In PV devices, photon energies greater than the cutoff (or band-gap) energy are dissipated as heat, while photons with wavelengths longer than the cutoff wavelength are not used.

The efficiency of a solar PV cell can be considered as the ratio of the electricity generated to the total, or global, solar irradiation. In this definition only the electricity generated by a solar PV cell is considered. Other properties of PV systems, which may affect efficiency, such as ambient temperature, cell temperature and chemical components of the solar cell, are not directly taken into account. Three PV system efficiencies can be considered: power conversion efficiency, energy efficiency and exergy efficiency. Energy (η) and exergy (ψ) efficiencies for PV systems are calculated as per below:

$$\eta = \text{energy in products/total energy input} \qquad (21)$$

$$\psi = \text{exergy in products/total exergy input} \qquad (22)$$

The actual output of the SPV module is defined as per Eq. (23):

$$Q_o = V_{oc} I_{sc} FF \qquad (23)$$

where V_{oc} is open circuit voltage, I_{sc} is short circuit current and FF is fill factor. The fill factor (FF) can be expressed as:

$$FF = \frac{V_m I_m}{V_{oc} I_{sc}} \qquad (24)$$

Using the above definition, Eq. (24) is expressed as:

$$Q_o = V_m I_m \qquad (25)$$

The energy efficiency is defined as [Pandey, 2013]:

$$\eta = \frac{V_{oc} I_{sc}}{I_s A} \qquad (26)$$

The input exergy i.e. exergy of solar radiation is given by:

$$Ex_{solar} = Ex_{in} = \left(1 - \frac{T_a}{T_s}\right) I_s A \qquad (27)$$

where T_s is the temperature of sun which is taken as 5777 K. The exergy output of the SPV systems can be given as follows:

$$Ex_{out} = Ex_{elec} + Ex_{therm} + E\dot{x}_d = E\dot{x}_{elec} + I' \qquad (28)$$

where $I' = \Sigma Ex = Ex_{d,elec} + Ex_{d,therm}$, which includes internal as well as external losses.

$$Ex_{elec} = E_{elec} - I' = V_{oc} I_{sc} - (V_{oc} I_{sc} - V_m I_m) \qquad (29)$$

where $V_{oc} I_{sc}$ stands for the electrical energy and $V_{oc} I_{sc} - V_m I_m$ stands for the electrical exergy destruction. Based on Eq. (29) the electrical exergy can be expressed as below:

$$Ex_{elec} = V_m I_m \qquad (30)$$

The thermal exergy of the system (Ex_{therm}), which is defined as the heat loss from the photovoltaic surface to the ambient is given by Eq. (31).

$$Ex_{therm} = \left(1 - \frac{T_a}{T_{cell}}\right) \dot{Q} \qquad (31)$$

where $\bar{Q} = h_{ca}A (T_{cell} - T_a)$ and $h_{ca}A = 5.7 + 3.8\,v$ are the heat transfer coefficients and v is the wind speed. Based on the above, the exergy output of a solar PV system is given by Eq. (32).

$$Ex_{pv} = V_m I_m - \left(1 - \frac{T_a}{T_{cell}}\right) h_{ca} A(T_{cell} - T_a) \tag{32}$$

The power conversion efficiency η_{pce} of a solar PV can be defined as the ratio of the actual electrical output to the input energy ($I_s A$) on the PV surface [Pandey, 2013]:

$$\eta_{pce} = \frac{V_m I_m}{I_s A} \tag{33}$$

The effect of temperature on the efficiency of the PV module can be obtained from the fundamental equations:

$$\eta_c = \eta_{Tref}\left[\beta_{ref}(T_c - T_{ref}) + \gamma \log_{10} I_s\right] \tag{34}$$

where η_{Tref} is the module's electrical efficiency at the reference temperature, T_{ref} and at solar radiation of 1000 W/m², β_{ref} is the temperature coefficient and γ is the solar radiation coefficient [Notton et al., 2005]. Based on Eq. (35) the temperature coefficient can be calculated:

$$\beta_{ref} = \frac{1}{T_o - T_{ref}} \tag{35}$$

In case there are variations in ambient temperature and irradiance the cell temperature can be estimated using the linear approximation [Luque and Hegedus, 2003]:

$$T_C = T_a + \left[\frac{T_{NOCT} - 20}{800\ \text{W}/\text{m}^2}\right] I_s \tag{36}$$

where T_{NOCT} is the Nominal Operating Cell Temperature (NOCT). Therefore efficiency can be given by Eq. (37):

$$\eta_c = \eta_{ref}\left[1 - \beta_{ref}\left[T_a - T_{ref} + (T_{NOCT} - 20)x\frac{I_s}{I_{S_{NOCT}}}\right] + \gamma \log_{10} I_s\right]100 \tag{37}$$

Usually $T_{ref} = 25\ °C$, average $\eta_{ref} = 12\%$ and average $\beta_{ref} = 0.0045$ K. Therefore, the exergy efficiency can be given as below [Pandey, 2013]:

$$\psi = \frac{V_m I_m - (1 - (T_a / T_{cell}))\, h_{ca} A(T_{cell} - T_a)}{(1 - (T_a / T_s))I_s A} \tag{38}$$

2.3.1.2 Case Studies

The physical and chemical principles of solar photovoltaic (SPV) conversion systems have been previously examined [Bisquert et al., 2004]. It has been indicated that the open-circuit voltage and chemical potential of a SPV cell depends on the Carnot and on statistical factors. In addition, energy and exergy analysis have been employed in India for a photovoltaic (PV) and a photovoltaic-thermal (PV/T) system [Joshi et al., 2009]. The energy efficiency was found to be between 33–45%, while the corresponding exergy efficiency was found to be between 11–16%. Nonetheless, the exergy efficiency for PV alone was calculated between 8–14% for a typical set of operating parameters.

Exergy analysis has also been employed for the evaluation of the performance of a solar photovoltaic thermal (PV/T) air collector [Sarhaddi et al., 2010]. An improved electrical model was used, in order to estimate the electrical parameters of the air collector; whereas a modified equation – taking into consideration design and climatic parameters was proposed.

The thermal, electrical, overall energy as well as exergy efficiency of PV/T air collector (taken into consideration sample climatic, operating and design parameters) were calculated as 17.18, 10.01, 45 and 10.75% respectively.

An extensive literature review on thermal modeling of photovoltaic (PV) modules and their applications indicated that the photovoltaic-thermal (PV/T) modules are of great significance and there is a great potential in improving their performances [Tiwari et al., 2011]. In this study it was shown that in regards to energy payback time (EPBT) and Energy Production Factor (EPF) the CIGS solar cells in the BIPVT system are the most suitable; whereas in regards Life Cycle Conversion Efficiency (LCCE) the building integrated photovoltaic thermal (BIPVT) system was the most suitable.

Energy and exergy analysis has been employed, in order to compare the performance of different types of SPV modules, such as monocrystalline silicon (m-Si), polycrystalline silicon (p-Si), amorphous silicon (a-Si), Cd-Te, CIGS and hetero junction with intrinsic thin layer (HIT) [Vats and Tiwari, 2012]. It was found that as the cell temperature increased the exergy efficiency decreased with maximum annual electrical energy produced by HIT to reach 810 kWh. Nonetheless, the maximum annual thermal energy produced by a-Si was 464 kWh; whereas the efficiency of HIT and Si module was found to be 16.0% for HIT and 6.0% for the a-Si respectively.

Energy and exergy analysis has also been carried out for a hybrid solar-fuel cell combined heat power system. Figure 2.2 illustrates the schematic diagram of the experimental set-up [Hosseini et al., 2013]. The energy and exergy efficiency of the PV system were found to be 17 and 18.3%, respectively. However, the total (PV and fuel cell combined) energy and exergy efficiencies were found to be 55.7 and 49% respectively.

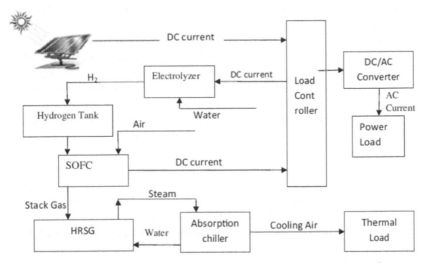

Fig. 2.2. Schematic diagram of the set-up [Hosseini et al., 2013]

2.3.2 Wind Energy Systems

Wind power is a form of renewable energy in that it is replenished daily by the sun. Warm air rises as portions of the Earth are heated by the sun, and other air rushes in to fill the low-pressure areas, creating wind power. Wind is slowed dramatically by friction as it flows over the ground and vegetation, often causing it not to be very windy at the ground level. Wind can be accelerated by major land forms, leading some regions to be very windy while other areas remain relatively calm. When wind power is converted to electricity, it can be transported over long distances and thus can serve the needs of urban centers where large populations live.

Wind energy is considered to be a renewable energy source of great significance. As a matter of fact, during the past decades the concerns over energy supply security in conjunction with the drive towards the creation of a sustainable energy future, fueled a lasting interest in this form of renewable energy sources. Recent technological developments indicate that wind generators have become a success story, being a mature, cost effective technology, with reasonable efficiency rates, high reliability and availability values [Viertel et al., 2005; Kaltschmitt et al., 2007]. During the past 20 years, technological developments have led to a significant increase of the nominal power of wind generators (by about two orders of magnitude). In addition, the continuous increase in the consumption of conventional energy resources have reduced the cost of generated energy and the wind energy industry has developed into a leading branch of the electricity generation section. It is of interest to note that 3 MW generators are already commercially viable systems, whilst generators with a nominal power rating of about 5 MW are in the final stages of trials [Krokoszinski, 2003]. Similar observations can be made to components like blades, power control, system of power transmission as well as concerning the increasing degree of wind parks penetration in not interconnected electrical systems and weak power grids (e.g. non-interconnected insular systems). Meteorological variables such as temperature, pressure and moisture play important roles in the occurrence of wind; given that pressure forces lead to kinetic energy that is observed as wind. These parameters can be described by expressions based on continuity principles, the first law of thermodynamics, Newton's law and the state law of gases.

Wind energy E is the kinetic energy of a flow of air of mass m at a speed v. The mass m is difficult to measure and can be expressed in terms of volume V through its density $\rho = m/V$. The volume can be expressed as $V = AL$ where A is the cross-sectional area perpendicular to the flow and L is the horizontal distance. Physically, $L = Vt$ and wind energy can be expressed as:

$$E = \tfrac{1}{2}\, p\, A_t V^3 \tag{39}$$

The retardation of wind passing through a windmill occurs in two stages: before and after its passage through the windmill rotor. Provided that a mass m is air passing through the rotor per unit time, the rate of momentum change is $m\,(v_1 - v_2)$ which is equal to the resulting thrust. v_1 and v_2 represent upwind and downwind speeds at a considerable distance from the rotor. The power absorbed P can be expressed as [Betz A. 1946]:

$$P = m\,(v_1 - v_2)\,\bar{V} \tag{40}$$

The rate of kinetic energy change in wind can be expressed as:

$$E_K = \tfrac{1}{2}\, m\,(v_1^2 - v_2^2) \tag{41}$$

The power extracted by the rotor is given by Eq. (42), as per below:

$$P = p\bar{A}\,V\,(v_1 - v_2)\,\bar{V} \tag{42}$$

Furthermore,

$$P = p\bar{A}\,V^2\,(v_1 - v_2) = p\,A\,(v_1 + v_2/2)^2\,(v_1 - v_2) \tag{43}$$

and

$$P = p\,A V_1^{\,3}\,[(1 + a)\,(1 - a^2)] \text{ where } a = v_2/v_1 \tag{44}$$

The power from wind's kinetic energy depends on the wind's velocity cube and the square of the turbine radius. This means that if the radius is doubled, the power of the wind will increase four times. The ratio between the rotor power and the power in the wind is expressed with the rotor power coefficient Cp:

$$Cp = \text{Rotor power}/\text{Power in the wind} \qquad (45)$$

A wind turbine operates at certain wind speeds and at a certain value (cut-in wind speed), no power is produced. The power generated by a wind turbine depends on both the design characteristics of the turbine and the properties of the wind resource. These parameters determine the capacity factor C_F (ratio of average power output to rated power of the turbine).

$$C_F = P_w/P_R \qquad (46)$$

The capacity factor significantly affects the energy production. Lenzen and Munksgaard [2002] have reported capacity factors ranging from 7.9 to 50.4%.

2.3.2.1 Exergy Analysis

The application of exergy analysis to wind energy systems has been investigated in many studies. These include the studies of Koroneos et al. [2003], who investigated the utilization of the wind's potential in relation to the wind speed and exergy losses in the different components of a wind turbine – i.e., rotor, gearbox and generator. In addition, Jia et al. [2004] took into consideration wind power for air compression systems operating over specified pressure differences. They estimated exergy components and presented pressure differences, by considering two systems (a wind turbine and an air compressor) as a united system. The wind speed thermodynamic characteristics have also been analyzed by using the cooling capacity of wind as a renewable energy source [Goff et al., 1999]. An exergy formulation for wind energy has also been suggested for the total exergy for wind energy [Sahin et al., 2006].

For a flow of matter at temperature T, pressure P, chemical composition μ_j of species j, mass m, specific enthalpy h, specific entropy s, and mass fraction x_j of species j, the specific exergy is expressed as:

$$e_x = [k_e + p_e + (h - h_0) - T_o(S - S_0)] + \left[\Sigma_j(\mu_{10} - \mu_{100})x_j\right] \qquad (47)$$

where h, k_e, p_e and e_x denote specific values of enthalpy, kinetic energy, potential energy and exergy, T_o, P_o and μ_{100} are intensive properties of the reference environment. The physical component (first term in square brackets on the right side of the above equation) is the maximum available work from a flow as it is brought to the environmental state. The chemical component (second term in square brackets) is the maximum available work extracted from the flow as it is brought from the environmental state to the dead state. For a wind turbine, kinetic energy is dominant and there is no potential energy change or chemical component. The exergy associated with work is expressed by following equation:

$$Ex^w = W \tag{48}$$

The exergy of wind energy can be estimated with the work exergy expression, because there are no heat and chemical components.

The energy (η) and exergy (ψ) efficiencies for the principal types of processes considered are based on the ratio of product to total input. Here, exergy efficiencies can often be written as a function of the corresponding energy efficiencies. The efficiencies for electricity generation in a wind energy system involve two important steps:

1. *Electricity generation from shaft work*: The efficiencies for electricity generation from the shaft work produced in a wind energy system are both equal to the ratio of the electrical energy generated to the shaft work input.

2. *Shaft work production from the kinetic energy of wind*: The efficiencies for shaft work production from the kinetic energy of a wind-driven system are both equal to the ratio of the shaft work produced to the change in kinetic energy Δke in a stream of matter *ms*.

The input and output variables for the system are described in Fig. 2.3. Output wind speed is estimated using the continuity equation. The total electricity generated is related to the decrease in wind potential. Subtracting the generated power from the total potential gives the wind turbine back-side wind potential:

$$V_2 = \sqrt[3]{\frac{2(E_{potential} - E_{generated})}{\rho At}} \tag{49}$$

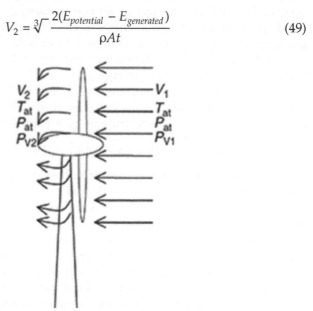

Fig. 2.3. Wind turbine and representative wind energy input and output variables [Anyanwu and Ogueke, 2005]

Total kinetic energy difference gives the generated electricity which can be written as:

$$\Delta KE = E_{\text{generated}} \tag{50}$$

Air mass flow with time depends on density and wind speed, and can be shown as:

$$\dot{m} = \rho A V \tag{51}$$

The exergy of a matter flow is defined as the maximum work that can be acquired when the air flows from state (T_2, P_2) to the ambient state (T_1, P_1). The enthalpy changes ΔH from state 1 and state 2 are expressed as:

$$\Delta H = \dot{m} C_p (T_2 - T_1) \tag{52}$$

where $\dot{m}$ is mass flow rate of air, which depends on time, T_1 is the wind chill temperature at the input to the wind turbine; and T_2 is the wind chill temperature at the exit of the wind turbine. The total entropy of the system and entropy difference can be written as:

$$\Delta S = \Delta S_{\text{system}} + \Delta S_{\text{surround}} \tag{53}$$

where

$$\Delta S = \dot{m} T_{\text{at}} \left(C_p \ln\left(\frac{T_2}{T_1}\right) - R \ln\left(\frac{P_2}{P_1}\right) \frac{Q_{\text{loss}}}{T_{\text{at}}} \right) \tag{54}$$

and

$$P_i = P_{\text{at}} \pm \frac{\rho}{2} V^2 \tag{55}$$

$$Ex = E_{\text{generated}} + \dot{m} C_p (T_2 - T_1) + \dot{m} T_{\text{at}} \left(C_p \ln\left(\frac{T_2}{T_1}\right) - R \ln\left(\frac{P_2}{P_1}\right) - \frac{Q_{\text{loss}}}{T_{\text{at}}} \right) \tag{56}$$

ΔS is the specific entropy change, T_{at} is the atmospheric temperature, P_2 is the pressure at the exit of the wind turbine for a wind speed V_2 and P_1 is the pressure at the inlet of the wind turbine for a wind speed V_1, Q_{loss} represents heat losses from the wind turbine and T average is the mean value of input and output wind chill temperatures. Thus, the total exergy for wind energy can be expressed using the above equations as:

$$Ex = E_{\text{generated}} + \dot{m} C_p (T_2 - T_1) + \dot{m} T_{\text{at}} \left(C_p \ln\left(\frac{T_2}{T_1}\right) - R \ln\left(\frac{P_2}{P_1}\right) - \frac{Q_{\text{loss}}}{T_{\text{at}}} \right) \tag{57}$$

2.3.2.2 Case Study: Application to Ios, Greece

In order to evaluate and assess wind energy potential, a database is considered of hourly wind speed and direction measurements of three stations in the island of Ios is taken into consideration. Ios (25.19°E longitude, 36.42°N latitude) is a Greek island in the Cyclades group in the Aegean Sea. Ios is a hilly island with cliffs down to the sea on most sides, situated halfway between Naxos and Santorini. It is about 18 kilometers long and 10 kilometers wide, with an area of about 109 square kilometers. Population was 2,024 in 2011 (down from 3,500 in the 19th century). This area comes under the influence of the mild Mediterranean climate during summer, and consequently experiences dry and hot spells for about 4 to 5 months, with comparatively little rainfall.

Ios is part of Paro-Naxia autonomous energy system, which includes Paros, Antiparos, Folegandros, Ios, Irakleia, Koufonisi, Naxos, Sikinos and Sxoinousa. The local power station of this system is located in Paros. The annual rate of increase of electricity demand comes up to 12%. Three wind turbines are in operation. Table 2.2a lists the nominal power outputs of the wind turbines for several wind speeds.

Table 2.2a. Technical characteristics of the three wind engines [NEG – Micon Technical Data]

Wind velocity/m s^{-1}	Wind turbine types		
	600 kW/48 m	*750 kW/48 m*	*1 MW/60 m*
6	93	95	150
7	153	168	248
8	235	259	385
9	329	362	535

The exergy of the wind turbine is going to be calculated by following equation:

$$e_{XWE} = \frac{1}{2} V_1^2 \tag{58}$$

In order to calculate the exergy efficiency of a wind energy system, consisting of three wind turbines, the following data are used: Air pressure is at 1 atm, air temperature is at 300 K, air density is 1.225 kg m^{-3}. The received power of the engine is given by following equation:

$$E_{in} = m e_{XWE} \tag{59}$$

where m is the air mass flow through the wind turbine. Based on the methodology presented in the previous section and taken into consideration the nominal power of the Neg-Micon wind engines, the exergy input, and the exergy efficiencies for the three wind turbines at different wind speeds are calculated. The results are presented in Table 2.2b.

Table 2.2b. Exergy efficiencies of three different wind turbines
[Koroneos et al., 2003]

Wind velocity/m s^{-1}	Wind turbine types		
	600 kW/48 m	750 kW/48 m	1 MW/60 m
6	38.8%	39.7%	40%
7	40.24%	44.19%	41.75%
8	41.41%	45.64%	43.42%
9	40.71%	44.6%	42.37%

2.4 Concluding Remarks

This chapter presented the application of exergy analysis to renewable energy systems. As it was shown, exergy is an effective tool in describing the use of energy resources and determine the future energy demand of an energy system. It can provide important knowledge and understanding to identify where large improvements could be obtained by applying efficient technologies and using more efficient energy resource conversions.

Exergy analysis can help to reach the objectives for the creation of a sustainable energy future by increasing the energy efficiencies of processes utilizing sustainable energy resources. The utilization of renewable energy offers a wide range of exceptional benefits including energy security, social and economic development, energy access, climate change mitigation and reduction of environmental and health impacts. The improvement of renewable energy technologies will assist the creation of a low carbon energy future, will enhance the road towards sustainability and provide a solution to several energy related environmental problems. In this sense, exergy analysis constitutes a suitable tool for governments and businesses for the optimization of renewable energy systems. The results of exergy analyses of processes and systems might have direct implications on application decisions and on research and development directions. Further, exergy analyses more than energy analyses provide insights into the "best" directions for further research.

Nomenclature

A – area
C – specific heat
Cp – specific heat at constant pressure
e – specific energy
E – energy; total solar energy reaching solar pond; wind energy
E potential – wind potential energy
E generated – electricity generated by wind turbine
ex – specific exergy
Ex – exergy
Ex_Q – exergy associated with heat Q
Ex_W – exergy associated with work W
F – absorbed energy percentage at δ-thickness region
h – specific enthalpy; ratio
H – enthalpy
I – number of the layers; irreversibility
k – thermal conductivity
ke – specific kinetic energy
KE – kinetic energy
m – mass
M – mass
N – number of moles
P – pressure
P_1 – pressure at inlet to wind turbine
P_2 – pressure at exit from wind turbine
pe – pecific potential energy
PE – potential energy
q – heat transfer per unit area
Q – heat
Q_r – heat transfer into system across region r on system boundary
R – thermal resistance
s – specific entropy
S – entropy; thickness
S_{gen} – entropy generation
ΔS – entropy change
t – time
T – temperature
T_{air} – air temperature
T_{windch} – wind chill temperature
u – specific internal energy
U – internal energy; heat loss from pond surface to air

v – specific volume
V – volume
V – wind speed
V_1 – upwind speed far from wind turbine rotor
V_2 – downwind speed far from wind turbine rotor
W – work; shaft work
x – mass fraction
Δx – thickness of horizontal layers
X – zone thickness
y – mole fraction
Ac – area of collector (m^2)
C – specific heat (kJ/kg K)
$\dot{E}$ – energy flow rate (kW)
$\dot{Ex}$ – exergy flow rate (kW)
$\dot{F}$ – exergy rate of fuel (kW)
h – specific enthalpy (kJ/kg)
I – current (Amps)
Is – intensity of solar radiation (W/m^2)
$\dot{I}$ – rate of irreversibility (kW)
L – latent heat of vaporization (J/kg)
M – moisture content
$\dot{m}$ – mass flow rate (kg/s)
$\dot{P}$ – exergy rate of the product (kW)
Pf – energy input to the fan (kWh)
P – pressure (N/m^2)
$\dot{Q}$ – heat transfer rate (kW)
R – ideal gas constant (kJ/kg K)
$\dot{S}$ – mass entropy rate at mass rate (kW/K)
$\dot{S}$ – heat entropy rate at heat rate (kW/K)
s – specific entropy (kJ/kg K)
t – time period (s)
ΔT – temperature difference
T – temperature (°C or K)
UL – overall heat loss coefficient (W/m^2K)
V – voltage (V)
$\dot{W}$ – work rate (kW)

Greek symbols
ε – exergy or second law efficiency
η – energy or first law efficiency
Ψ – flow exergy (kJ/kg)
δ – fuel depletion rate
Δ – interval
χ – relative irreversibility

ξ – productivity lack
β – packing factor of solar cell
ρ – density (kg/m^3)
f – exergetic factor, product

REFERENCES

Ahrendts, J. (1980). Reference states. Energy – Int J 5: 667–678.

Anyanwu, E.E. and Ogueke, N.V. (2005). Thermodynamic design procedure for solid adsorption solar refrigerator. Renew Energy 30: 81–96.

Badescu, V. (2014). Is Carnot efficiency the upper bound for work extraction from thermal reservoirs. Europhysics Letters 106(1).

Baehr, H.D. and Schmidt, E.F. (1963). Definition und Berechnung von Brennst off exergien (Definition and calculation of fuel exergy). Brennst–Waerme–Kraft 15: 375–381.

Beccali, G., Cellura, M. and Mistretta, M. (2003). New exergy criterion in the multi-criteria context: A life cycle assessment of two plaster products. Energy Conversion Management 44: 2821–2838.

Beretta, G.P. and Gyftopoulos, E.P. (2015). Electromagnetic radiation: A carrier of energy and entropy. Journal of Energy Resources Technology, March 2015 vol. 137/021005-1. ASME.

Betz A. (1946). Windenergie und ihre Ausnutzung durch Windmühlen. Gttingen: Vandenhoek and Ruprecht, Göttingen.

Bisquert, J., Cahen, D., Hodes, G., Ruhle, S. and Zaban, A. (2004). Physical chemical principles of photovoltaic conversion with nanoparticulate, mesoporous dye-sensitized solar cells. Journal of Physical Chemistry B 108: 8106–8118. doi: 10.1021/jp0359283

Bosnjakovic, F. (1963). Bezugszustand der Exergie eines reagiernden Systems (Reference states of the exergy in a reacting system). Forsch. Ingenieurw 20: 151–152.

Boyle, G. (2004). Renewable Energy: Power for a Sustainable Future. Second edition. Oxford University Press.

Brodyanski, V.M., Sorin, M.V. and Le Goff, P. (1994). The Efficiency of Industrial Processes: Exergy Analysis and Optimization. London: Elsevier.

Caliskan, H. (2015). Novel approaches to exergy and economy based enhanced environmental analyses for energy systems. Energy Conversion Management 89: 156–161.

Candau, Y. (2003). On the exergy of radiation. Solar Energy. 75: 241–247.

Chen, G. and Chen, B. (2009). Extended-exergy analysis of the Chinese society. Energy 34: 1127–1144.

Cornelissen, R.L. (1997). Thermodynamics and Sustainable Development: The Use of Exergy Analysis and the Reduction of Irreversibility. Ph.D. Thesis, University of Twente, Enschede, The Netherlands.

Crane, P., Scott, D.S. and Rosen, M.A. (1992). Comparison of exergy of emissions from two energy conversion technologies, considering potential for environmental impact. International Journal of Hydrogen Energy 17: 345–350.

DiPippo, R. (2004). Second Law assessment of binary plants generating power from low-temperature geothermal fluids. Journal of Geothermics, Elsevier 33: 565–586.

Dupeyrat, P., Menezo, C. and Fortuin, S. (2014). Study of the thermal and electrical performances of PVT solar hot water system. Energy and Buildings, Elsevier 68(Part C): 751–755.

EC (2015). Communication from the Commission to the European Parliament, the Council, the European Economic and Social Committee, the Committee of the Regions and the European Investment Bank, A Framework Strategy For A Resilient Energy Union With A Forward-Looking Climate Change Policy.

Fudholi, A., Sopian, K., Othman, M.Y., Ruslan, M.H. and Bakhtyar, B. (2013). Energy analysis and improvement potential of finned double-pass solar collector. Energy Conversion and Management 75: 234–240. https://doi.org/10.1016/j.enconman.2013.06.021

Fudholi, A., Sopian, K., Yazdi, M.H., Ruslan, M.H, Gabbasa, M. and Kazem, H.A. (2014). Performance analysis of solar drying system for red chili. Sol. Energy 99.

Gaggioli, R.A. and Petit, P.J. (1977). Use the second law first. Chemtech. 7: 496–506.

Gaggioli, RA. (1983). Second law analysis to improve process and energy engineering. pp. 3–50. *In*: Efficiency and Costing: Second Law Analysis of Processes. ACS Symposium Series 235, Washington, DC: American Chemical Society.

Gasparatos, A., El-Haram, M. and Horner, M.A. (2009). Longitudinal analysis of the UK transport sector, 1970-2010. Energy Policy 37(2): 623–632.

Ge, Z., Wang, H., Wang, H., Zhang, S. and Guan, X. (2014). Exergy analysis of flat plate solar collectors. Entropy 16(5): 2549–2567.

Goff, L.H., Hasert, U.F. and Goff, P.L. (1999). A "new" source of renewable energy: The coldness of the wind. Revue Generale de Thermique 38(10): 916–924.

Gomri, R. (2013). Simulation study on the performance of solar/natural gas absorption cooling chillers. Energy Conversion and Management 65: 675–681.

Gunnewiek, L.H. and Rosen, M.A. (1998). Relation between the exergy of waste emissions and measures of environmental impact. International Journal of Environmental Pollution 10(2): 261–272.

Hammond, G.P. and Stapleton, A.J. (2001). Exergy analysis of the United Kingdom energy system. Proc. Inst. Mech. Eng. Part A J. Power Energy 215: 141–162.

Hepbasli (2008). A key review on exergetic analysis and assessment of renewable energy resources for a sustainable future. Renewable and Sustainable Energy Reviews 12: 593–661.

Hermann, W.A. (2006). Quantifying global exergy resources. Energy 31(12): 1685–1702.

Hosseini, M., Dincer, I. and Rosen, M.A. (2013). Hybrid solar e fuel cell combined heat and power systems for residential applications: Energy and exergy analyses. Journal of Power Sources 221: 372–380.

Hou, H., Yu, Z., Yang, Y., Zhou, C. and Song, J. (2014). Exergy analysis of parabolic trough solar collector. Taiyangneng Xuebao/Acta Energiae Solaris Sin 35(6): 1022–1028.

Jaber, J.O., Al-Ghandoor, A. and Sawalha, S.A. (2008). Energy analysis and exergy utilization in the transportation sector of Jordan. Energy Policy 36(8): 2995–3000.

Ji, X. and Chen, G.Q. (2006). Exergy analysis of energy utilization in the transportation sector in China. Energy Policy 34(14): 1709–1719.

Jia, G.Z., Wang, X.Y. and Wu, G.M. (2004). Investigation on wind energy–compressed air power system. Journal of Zhejiang University Sciences 5(3): 290–295.

Joshi, A.S., Dincer, I. and Reddy, B.V. (2009). Thermodynamic assessment of photovoltaic systems. Solar Energy 83(8): 1139–1149.

Kaltschmitt, M., Streicher, W. and Wiese, A. (2007). Renewable Energy, Technology, Economics and Environment. ISBN 978-3-540-70947-3. Springer Berlin, Heidelberg, New York.

Kasaeian, A.B., Mobarakeh, M.D., Golzari, S. and Akhlaghi, M.M. (2013). Energy and exergy analysis of air PV/T collector of forced convection with and without glass cover. International Journal of Engineering, Transactions B: Applications 26(8): 913–926.

Koroneos, C., Spachos, N. and Moussiopoulos, N. (2003). Exergy analysis of renewable energy sources. Renewable Energy 28: 295–310.

Koroneos, C., Nanaki, E. and Xydis, G. (2011). Exergy analysis of the energy use in Greece. Journal of Energy Policy, Elsevier Publishers 39: 2475–2481.

Koroneos, C. and Nanaki, E. (2008). Energy and Exergy Utilization Assessment of the Greek Transport Sector. Journal of Resources, Conservation and Recycling, Elsevier 52(5): 700–706.

Koroneos, C., Spachos, T. and Mousiopoulos, N. (2003). Exergy analysis of renewable energy sources. Journal of Renewable Energy 28: 295–310.

Kotas, T.J. (1995). The Exergy Method of Thermal Plant Analysis. Reprint edn. Malabar, Florida: Krieger.

Krokoszinski, H.J. (2003). Efficiency and effectiveness of wind farms—keys to cost optimized operation and maintenance. Renew Energy 28(14): 2165–2178.

Landsberg, P.T. and Mallinson, J.R. (1976). Thermodynamic constraints, effective temperatures and solar cells. pp. 27-35. *In*: Coll. Int. sur l'Electricite Solaire, CNES, Toulouse.

Lenzen, M. and Munksgaard, J. (2002). Energy and CO_2 life cycle analyses of wind turbines—review and applications. Renewable Energy 26(3): 339–362.

Luque, A. and Hegedus, S. (2003). Handbook of Photovoltaic Science and Engineering. The Atrium, Southern Gate. Chichester, West Sussex PO 198SQ. England: John Wiley & Sons Ltd.

Madadi, V., Tavakoli, T. and Rahimi, A. (2014). First and second thermodynamic law analyses applied to a solar dish collector. Journal of Non-Equilibrium Thermodynamics 39(4): 183–197.

NEG-Micon Technical Data: http://www.windpower.dk/tour/wres/pow/index.htm.

Nematollahi, F., Rahimi, A. and Gheinani, T.T. (2013). Experimental and theoretical energy and exergy analysis for a solar desalination system. Desalination 317: 23–31.

Notton, G., Cristofari, C., Mattei, M. and Poggi, P. (2005). Modelling of a double-glass photovoltaic module using finite differences. Applied Thermal Engineering 25: 2854–2877.

Oztop, H.F., Bayrak, F. and Hepbasli, A. (2013). Energetic and exergetic aspects of solar air heating (solar collector) systems. Renewable and Sustainable Energy Reviews 21: 59–83.

Ozturk, H.H. (2004). Experimental determination of energy and exergy efficiency of the solar parabolic-cooker. Solenergy 77: 67–71.

Padilla, R.V., Fontalvo, A., Demirkaya, G., Martinez, A. and Quiroga, A.G. (2014). Exergy analysis of parabolic trough solar receiver. Journal of Applied Thermal Engineering 67(1–2): 579–586.

Pandey, A.K. (2013). Exergy Analysis and Exergoeconomic Evaluation of Renewable Energy Conversion Systems. Ph.D. Thesis. School of Energy Management, Shri Mata Vaishno Devi University, Katra, India.

Petela, R. (2003). Exergy of undiluted thermal radiation. Solar Energy 74: 469–488.

Petela, R. (2005). Exergy analysis of the solar cylindrical-parabolic cooker. Solar Energy 79: 221–233.

Press, W.H. (1976). Theoretical maximum for energy from direct and diffuse sunlight. Nature 264: 734–735.

Rodriguez, L.S.J. (1980). Calculation of available-energy quantities. pp. 39–60. *In*: Thermodynamics: Second Law Analysis. ACS Symposium Series; 122.

Rosen, M.A. and Dincer, I. (1997). Sectoral energy and exergy modeling of Turkey. ASME Journal of Energy Resources Technology 119: 200–204.

Rosen, M.A. and Dincer, I. (2004). A study of industrial steam process heating through exergy analysis. International Journal of Energy Research 28(10): 917–930.

Rosen, M.A., Le, M.N. and Dincer, I. (2005). Efficiency analysis of a cogeneration and district energy system. Journal of Applied Thermal Engineering 25(1): 147–159.

Rosen, M.A., Tang, R. and Dincer, I. (2004). Effect of stratification on energy and exergy capacities in thermal storage systems. International Journal of Energy Research 28: 177–193.

Rovira, A., Montes, M.J., Varela, F. and Gil, M. (2013). Comparison of heat transfer fluid and direct steam generation technologies for integrated solar combined cycles. Journal of Applied Thermal Engineering 52.

Sahin, A.D., Dincer, I. and Rosen, M.A. (2006). Thermodynamic analysis of wind energy. International Journal of Energy Research 30(8): 553–566.

Sarhaddi, F., Farahat, S., Ajam, H. and Behzadmehr, A. (2010). Exergetic performance assessment of a solar photovoltaic thermal (PV/T) air collector. Energy and Building 42: 2184–2199.

Skoglund, A., Leijon, M., Rehn, A., Lindahl, M. and Waters, R. (2010). On the physics of power, energy and economics of renewable electric energy sources – Part II. Renewable Energy 35(8): 1735–1740.

Strupczewskim, A. (2003). Accident risks in nuclear-power plants. Applied Energy 75(1–2): 79–86.

Sun, F., Zhou, W., Ikegami, Y., Nakagami, K. and Su, X. (2014). Energy, exergy analysis and optimization of the solar-boosted Kalina cycle system 11 (KCS-11). Renewable Energy 66.

Sussman, M.V. (1980). Steady-flow availability and the standard chemical availability. Energy – International Journal 5: 793–804.

Szargut, J. (1967). Grenzen fuer die Anwendungsmoeglichkeiten des Exergiebegriffs (Limits of the applicability of the exergy concept). Brennst.–Waerme–Kraft 19: 309–313.

Szargut, J. (2005). Exergy Method: Technical and Ecological Applications. Southampton, Boston: WIT Press.

Szargut, J., Morris, D.R. and Steward, F.R. (1988). Exergy Analysis of Thermal, Chemical, and Metallurgical Processes. Hemisphere Publishing Corporation.

Texas Renewable Energy Industries Alliance – http://www.treia.org (accessed online March 2018).

Tiwari, G.N., Mishra, R.K. and Solanki, S.C. (2011). Photovoltaic modules and their applications: A review on thermal modeling. Applied Energy 88: 2287–2304.

UNEP (2015). Green energy choices: The benefits, risks and trade-offs of low-carbon technologies for electricity production—Summary for policy makers. United Nations Environment Programme.

Van Gool, W. (1997). Energy policy: Fairy tales and factualities. pp. 93–105. *In*: Soares, O.D.D., da Cruz, A.M., Pereira, G.C., Soares, I.M.R.T., Reis, A.J.P.S. (Eds.). Innovation and Technology—Strategies and Policies. Kluwer: Dordrecht, The Netherlands.

Vats, K. and Tiwari, G.N. (2012). Energy and exergy analysis of a building integrated semi-transparent photovoltaic thermal (BISPVT) system. Applied Energy 96: 409–416.

Viertel, R., Kaltschmitt, M. and Tetzlaff, G. (2005). 3.000 bis 12.000 Quadratkilometer für offshore parks. pp. S.27–31. *In*: Erneuerbare Energien 3/2005 [in German].

Wall, G. (1990). Exergy conversion in the Japanese society. Energy 15: 435–444.

Wall, G., Sciubba, E. and Naso, V. (1994). Exergy use in the Italian society. Energy 19: 1267–1274.

Wang, J., Yan, Z., Zhao, P. and Dai, Y. (2014). Off-design performance analysis of a solar powered organic Rankine cycle. Energy Conversion Management 80.

Wepfer, W.J., Gaggioli, R.A. and Obert, E.F. (1979). Proper evaluation of available energy for HVAC. ASHRAE Transactions 85(1): 214–230.

Xydis, G., Koroneos, C. and Nanaki, E. (2011). Exergy based comparison of two Greek industries. International Journal of Exergy 8(4): 460–476.

Yang, L., Zhang, X. and Xu, G. (2014). Thermal performance of a solar storage packed bed using spherical capsules filled with PCM having different melting points. Energy Build. 68(Part B).

Zhang, X., Zhao, X., Shen, J., Xu, J. and Yu, X. (2014). Dynamic performance of a novel solar photovoltaic/loop-heat-pipe heat pump system. Journal of Applied Energy 114.

Electricity Markets and Renewable Energy Sources – A Smart City Approach

G. Xydis[1] and E. Nanaki[2]

[1] Department of Business Development and Technology, Aarhus University, Birk Centerpark 15, 7400 Herning, Denmark
[2] University of Western Macedonia, Department of Mechanical Engineering, Bakola & Sialvera, Kozani 50100, Greece

3.1 Introduction

A major transformation of the European energy system is currently taking place. The massive implementation of Renewable Energy Sources (RESs) and Distributed Energy Resources (DERs) needs to be supported by complimentary technologies for intelligent control. Technologies in the form of Smart grids, Smart cities and Smart building systems are required to maintain or increase system robustness, stability and security. Urban areas are in particular interesting due to their power dynamics and due to their high power and energy density. Urban areas account for approximately 70% of the primary EU energy consumption and the electricity share is expected to increase significantly with the integration of electric vehicles. E-mobility in urban areas enhances the effect of power dynamics and energy density.

There is currently a lack of understanding of the aggregated impact of local energy systems and components. How is the overall performance of the total energy system? Researchers, urban planners and technology developers require appropriate methods and tools for analyzing, visualizing and optimizing the performance of local energy systems based predominantly on aggregated DER and intermittent RES.

However, without a holistic research approach, the substantial energy, emissions and cost savings achievable by integrating the electricity, heating, cooling, water and transport systems cannot be accessed, and a fully renewable energy system will be difficult to be achieved. The required focus on energy systems integration calls for advanced ICT solutions, and a broader view of the smart grids concept.

The leading position of European academia and industry and the rapidly growing market for smart energy solutions indicates substantial scope for increased competitiveness and job creation within the smart cities field. There is a need for urban project developers to pioneer this research area, building short term operational integrated energy system models that feed longer term planning models, considering the spatio/temporal variations, interactions, dynamics and stochastics in the urban environment, and the need from production to consumption.

Cities, with a high density of energy activity and networks, offer the greatest potential for flexibility at the least cost, and play a dominant role in current and future energy consumption, making them an ideal basis for research on the way to a fossil-free future. Urban focused decision support tools should be developing to always keep stakeholders informed of the impact of cities' investments, and identify further opportunities for a modern energy system. For a fully renewable energy system worldwide [100% Renewable Energy, 2018], an overhaul of the operation, monitoring and planning of the entire energy system is needed [Føyn et al., 2011; Koroneos et al., 2010; Piwko et al., 2012].

3.1.1 Market Needs

Currently, numerous efforts are carried out globally for the development of energy-saving and Green-House Gas (GHG) emission reducing systems. Scientists' and decision makers' efforts intensively address one of the major power consuming areas during the last decades; namely, the energy usage in commercial buildings in developed countries. The majority of energy consumption addressed is in the form of electrical energy, as well as the energy used for heating which alone accounts for 70% of the total energy consumption within the urban environment. The way we deal with it is currently not smart at all. However, we are working on that.

The nature of this consumption is characterized by daily variations and cycles, which is in addition affected by stochastic fluctuations in weather. At the same time, there is an urgent need to/for increase in the utilization of renewable energy sources. Matching those two factors is an added challenge for the introduction of flexible and intelligent combinatorial scheduling, that will improve energy efficiency and reduce overall consumption through local synergies and smart integration of novel technologies in both existing and newer buildings and neighborhoods.

With the European Directive for energy performance in buildings, EU Member states are confronted with a challenge to meet the significant targets. Article 9.1. regulates that "Member States shall ensure that by 31 December 2020, all new buildings are nearly zero-energy buildings (1a) and after 31 December 2018, new buildings occupied and owned by public authorities are Nearly Zero Energy Buildings (NZEB). NZEBs have a very high energy performance and a major part of the amount of energy that these buildings require comes from renewable energy sources [Towards nearly zero energy buildings, 2013].

Via the smart cities projects all around the world there are those that work on topics of great potential, such as the energy positive neighborhood initiative, which has outlined conceptual ideas and will establish the framework for exploiting optimal energy synergies at neighborhood levels in the near future scenarios. Some of those are:

- Buildings with excess heat sharing with those short in demand
- Achieving energy efficient synergies by combining resources across private, public, and industrial actors
- Making non-traditional use of different types of spaces for example, using roads, parking spaces, etc., for communal gathering
- Enhancing integration of local renewable energy sources in rural and urban settings, and
- Increasing energy awareness and strengthening natural drivers/incitements for optimizing energy use.

In addition, there is the need to utilize state-of-the-art enabling technologies such as those available within ICT like smart grids and smart meters, local thermal and electrical renewable sources, ground-source heat, electrical vehicles, energy storages and two-way sharing networks and DC microgrids.

3.1.2 Energy Efficient Urban Neighborhoods

In a concept on Energy Efficient Urban Neighborhoods, the synergies achievable within city and urban neighborhoods using the strategic combination of thermal and electrical technologies via local microgrids should be explored.

A neighborhood energy management framework (i. Identification of local energy neighborhood areas, ii. technology guidance, and iii. modeling of energy savings by technology assessments with standardization and characterization of new technologies and construction area) could include the advanced use of ICT information management in orchestrating the intelligent interplay of existing and emerging energy saving technologies, as well as exploring the energy synergies at a neighborhood community level encompassing existing and future buildings and spaces

(Neighborhood energy management systems for daily operation of the system – mainly through an integration of existing tools). The idea is to explore the use of ICT to enable and optimize the demand-side response and control future energy efficient neighborhoods by:

- Achieving high overall reduction of energy consumption, or even negative consumption.
- Better energy efficiency via utilization and sharing of local production; i.e. achieving Synergies (City district energy modeling for thermal and electrical consumption and microgrids).
- Reaching higher utilization of external renewable energy with variable tariffs and smart grid technologies.
- Enabling energy security technologies [Chalvatzis and Rubel, 2015].

At the neighborhood level the synergies will have to be optimized by use of the combination of areas, facilities and technologies as in:

1. local renewable sources (solar, ground source heating, local wind power);
2. local energy storages and energy buffers (hot water tanks, batteries, and thermal capacities, e-vehicles) and
3. local energy sharing networks being thermal or electrical and smart grids, two-way energy flows.

Decision makers are focusing to advanced and integrated frameworks that will include toolboxes to be used by several actors consisting of:

- City planners and administrators – for making plans of new city areas and buildings; renovation and refurbishment strategies; that are truly low energy or energy positive areas and for achieving the best energy synergies and most sustainable energy profile of new city areas.
- Citizens and potential energy-neighborhood participants – seeking to identify and possible neighborhood synergies.
- Academics/Researchers – for further optimization and development of improved energy effective strategies and new technologies.
- Factories/Industries – for the identification of needs and markets for novel technologies which can be used at neighborhood/district levels.

The idea is to end up to a framework/guidance system that will have some of the features and functionalities below:

- Be able to make suggestions and give guidance for the implementation technologies/new installations at a neighborhood level which in combination will give the optimal energy efficiency and cost savings.
- The system will use analysis of city/rural maps/3D city models, satellite data, GEOS data to automatically identify synergy neighborhoods (e.g. by identification parameters such as building

types, free land spaces, free roof space, types of land use (e.g. unused areas, roads, parkings etc) and activities [Keramitsoglou et al., 2016]

- Be able to estimate energy/cost benefits achieved through the neighborhood level implementation of given green technologies, including electrical, thermal energy flows and networks, as well as the combination of the two (as in heat pumps).
- Be able to simulate energy profiles based on typical energy profiles for different types of buildings and spaces [Yu et al., 2017]
- Be able to provide a holistic approach which includes all elements of a city, e.g. buildings, roads, parking areas, recreational areas, etc.
- Can typically work on smaller neighborhoods in the size of 10-30 units (such as residential buildings, apartment blocks, roads and spaces, schools etc.)
- Be able to provide graphical 3D user interface with visual representation and user friendly control of steep learning curve.
- Multiuser support, definition of members and groups
- Emphasis on data security. Privacy and data encryption, data backup, synchronization, and reliability
- Means to define own and shared devices: Architecture allowing easy integration, plug-and-play, of new devices and systems
- Energy management optimized for the effectiveness of combined energy devices: buffers, energy sources, and sinks covering the whole neighborhood community.
- Enhanced predictive algorithms, with the use of Internet resources with local current and weather forecast and climate conditions as well as current and predicted renewable production and tariffs
- Learning system based on knowledge of past history
- High level of Interoperability with existing devices energy management systems. Operate as a monitoring layer in parallel with other smart grid functions.
- Develop business concepts of a local economy structure [Korhonen et al., 2018]
- Cloud data management and Internet of Things (IoT) platforms, for easy access, maintenance, speed and reliability.

3.1.3 Progress Beyond Network Synergies

Today many new technologies are being developed, however, their interdependent synergy and combination is a less investigated area, even though, the total energy saving impacts will be much higher. A good example is converting renewable electricity excess into thermal or cooling energy [Xydis, 2013a]. Another example is natural cooling of PV surfaces and increase in productivity [Xydis, 2013b].

Up to today energy systems optimization happened from the supply side and high voltage level grids that included centralized power plants.

By deployment of renewable energy technology and decentralized energy generation it is expected that soon 75% of energy will come from these sources so it is important to focus on system modeling at low and medium voltage level and include all possible energy flows and energy storages integration (e.g. microCHP, tri-generation, electric vehicles). Furthermore, it is important to measure all supply and demand in real time in order to have proper automation and control of the systems. Thus, ICT is playing the major role in modeling, optimization and control of local energy systems.

3.1.3.a Buildings Synergies

Zero Energy Buildings (ZEB) are in the family of green buildings or sustainable architectures, in the category of "net off-site zero energy use" or "Off-the-grid standalones". The prime aim is to be highly independent from the energy grid supply.

Through highly energy efficient technologies, including on-site energy harvesting, and a combination of energy producing technologies like Solar and Wind, and extremely efficient HVAC and Lighting technologies a significant reduction of energy consumption is achieved. The zero-energy design principle is becoming more practical to adopt due to the increasing costs of traditional fossil fuels.

ZEB are often designed to make dual use of energy including white goods; for example, using refrigerator exhaust to heat domestic water, ventilation air and shower drain heat exchangers, office machines and computer servers, and body heat to heat the building. These buildings make use of heat energy that conventional buildings may exhaust outside. They may use heat recovery ventilation, hot water heat recycling, combined heat and power, and absorption chiller units. Zero energy building employs carefully optimized energy management strategies, typically combine time tested passive solar, or natural conditioning, principles that work with the on-site assets and occupant behavior. Sunlight and solar heat, prevailing breezes, and the cool of the Earth below a building, can provide daylight and stable indoor temperatures with minimum mechanical means.

One of the main challenges is overcoming initial costs, and problems with optimized thermal envelope the embodied energy, heating and cooling energy and resource usage is higher than needed. ZEB by definition do not mandate a minimum heating and cooling performance level, thus allowing oversized renewable energy systems to fill the energy gap. There is a need to optimize combinations of RES technologies extending beyond the individual buildings and neighborhoods.

The way ATES (Aquifer Thermal Energy Storage) or UTES (Underground Thermal Energy Storage) works for cooling or low temperature heating is that energy is stored in aquifers 20 to 300 m below the surface. Water is pumped up from an aquifer, heat is extracted and relatively cold water is re-infiltrated in the same aquifer. A cold groundwater zone (5 to 10°C) develops around the infiltration well, which can be used for cooling in the summer. This water then absorbs heat from the building and energy can be stored in a corresponding warm water zone (15 to 30°C) and used to heat the building in winter. For heating in general a heat pump (not shown) will be used to boost the temperature in the building's loop. This water then absorbs heat from the building and energy can be stored in a corresponding warm water zone (15 to 30°C) and used to heat the building in winter. For heating in general a heat pump will be used to boost the temperature in the building's loop. This concept is a proven technology which has been applied in more than 1500 buildings in the Netherlands and also in some other countries. The potential of this technology in Europe is big. In order to apply this system the hydrogeological situation should be good.

The ATES technology can be combined in a heating and cooling grid in order to shave the peaks in energy demand and to store the surplus of heat and cold and release it at the right moments. By using ATES in a grid additional savings can be obtained on the energy bill and on CO_2, SO_4 and NO_x. On the grid are connected buildings which have a surplus of heat which can be used by buildings which have a heating demand. Cooling can be stored in the winter period and provided during the summer. ATES is doing the work of the balancing component in the grid.

3.1.3.b Microgrids Synergies

With the growth of Distributed Energy Resources (DERs), the need to mitigate the impact of volatile generation on the distribution networks should be met by implementing control algorithms to coordinate the participation from DERs within the urban environment. Microgrid is proposed to integrate DERs as one entity to distribution network, which comprises DERs together with storage devices at low voltage levels connected to distribution systems [Tsikalakis and Hatziargyriou, 2008].

Typical microgrid components include both electrical and heating combined heating and power plants (CHPs), solar photovoltaic (PV) modules, small Wind Turbines (WTs), electric vehicles, other small renewable generating units, storage devices, heating system, and various types of controllable loads. Storage devices are necessary due to different heat and electrical loads profiles [Marnay et al., 2008], while electric heater [Hernandez-Aramburo et al., 2005] is another key component which links the electricity and heat systems and provides flexibility to the system operation.

The difficulty of renewable generation forecasting, due to its intrinsic dependence on varying weather conditions, may impose significant challenges to the power system operation. The operation of DERs in the network can provide distinct benefits to the overall system performance, if managed and coordinated properly. The concept of microgrid is adopted in this project to optimally accommodate these energy carriers at different neighborhood levels, to provide cost minimum operations.

It is proven that renewable technologies fare surprisingly well in several electrification configurations. In addition to the proven economical values in the expensive off-grid category, they are also more economical in mini-grid applications even when compared with small grid-connected generation (less than 50 MW). Since power system planners generally operate on an incremental basis, with new capacity additions (generation, transmission or distribution) timed and sized to accommodate the location and pace of load growth, the findings here suggest that scale and insensitivity to fuel and technology change factors could affect the economics of choosing generation configurations in future.

Local micro wind turbines can also be bought for energy ratings of 500 W to 20 kW. Disadvantages of such systems can be the noise generation, high investment costs, need for relatively high wind speeds, variable energy output, and highly visual impact on the surroundings and neighbors. Photovoltaic solar panels have been experiencing a very high increase of utilization within cities. The efficiency has recently been improved making the systems more affordable.

Predictive methods and combination with energy storages and synergies with other RES technologies will be the key to maximized utilization of PV technology furthermore in the urban environment. There are several identified technologies which are expected in the future to contribute as local electrical energy storages and buffers even further.

Electric Vehicles (EVs) have high power and energy content due to battery technology and the bidirectional control of the charging process. Self-driving electric vehicles cars are gaining more and more attention and have started meeting grid needs as EVs do besides increased awareness and safety requirements. For instance, self-driven Waymo cars have completed 10, million miles (update: Oct., 2018) and became the most experienced driver ever (Fig. 3.1) [Waymo, 2018]. Furthermore, utilities with lower capacity like various energy carriers, such as refrigerators, freezers [Xydis, 2013a] and heat pumps, etc, which may still be added up to maximally utilize the energy carrying capability.

3.2 City District Energy Modeling and Electricity Markets

City modeling tools are seen as very strong for city planning and administration. They are typically versatile 3D platforms which could be employed into given specified use-case, being traffic, waste flow,

Fig. 3.1. Waymo: The most experienced driver

maintenance, energy estimates, water consumption, rain flooding etc. As these 3D city models get more detailed, more applications are appearing.

3.2.1 3D Modeling and Energy Management

There are currently several on-going efforts within 3D modeling of energy in buildings, districts and cities. Due to improved tools for the design and acquisition of 3D models, to the wider acceptance of 3D technology, and 3D spatial databases the creation and management of urban information spaces representing entire cities in the virtual world is feasible nowadays. There are recent efforts on developing even digital twins to simulate the urban environment which, of course, requires huge amount of data storage and processing [Mohammadi and Taylor, 2018].

Various modeling approaches are able to provide adequate accuracy to analyze the relationship of buildings and building energy systems with energy use in a district [Best et al., 2015; Liaros et al., 2016; Xydis, 2012a; Fuchs et al., 2016]. The urban information space is not limited to 3D building geometry but includes building semantics as well providing necessary information for urban planning, construction and management. However, sharing information between professionals from various disciplines and non-professional users such as citizens affected by planning proposals is a big challenge for the future.

Currently, traditional disciplines such as architecture, civil engineering and GIS create classic information islands with different focuses. Usually GIS is used to represent the current status of a whole city for administrative purposes. In contrast, planning and construction professionals focus on the future shape and status of a relatively small part of the city or just individual buildings in very great detail. In general, three categories of urban geo-information can be distinguished:

- GIS: two-dimensional maps, digital terrain model, 3D buildings with simple geometry of the entire city
- CAAD: detailed building models including interior
- BIM: building information model including information about structure and usage of individual buildings and their interior up to urban quarters.

Due to the historic development of domain-specific applications, systems and formats, the different data sources are not interoperable per se. Döllner and Hagedorn [2007] have shown an integration of these data on the visualization level using a service-based 3D viewer. However, new challenges such as sustainable development and energy-efficiency do not only require an integration on visualization level but an integration and convergence on data model level to enable intelligent data processing on the city scale level. Integration and convergence on data model level is a key enabler to simulate an entire urban environment. The ability to simulate entire urban environments has implications for urban planning including the ability to simulate urban redevelopment projects, noise pollution, daylight simulation, wind modeling, emergency planning, and energy-management.

Urban quarters generally lack efficient energy management, which causes primary energy losses of up to 30% of the actual energy efficiency potential of the urban site. A reduction of the CO_2 emissions of urban quarters of only 10% (approx. 150 million tons) would mean an enormous impact on EU greenhouse gas avoidance efforts. The inefficient urban energy management is mainly caused by a missing integration platform, where data from the demand side, such as commercial, residential or public buildings (exergetic performance, energy consumption, user behavior etc.) and from the supply side (district heating systems, district cogeneration, renewable energy systems, etc) can be collected and implemented in an efficient urban site energy management system.

To achieve the integration a 3D spatial data infrastructure has to be built in order to connect the existing information datasets. The core component of this data infrastructure is a data model that specifies the knowledge stored in the overall city information model. 3D city models are already used for automatic extraction of parameters for heat demand simulations, PV-Potential analysis, siting of renewables and the calculation and visualization of energy scenarios. The basis CityGML – which is an XML-based format open data model for the storage and exchange of virtual 3D city models – allows extending the model by further energy relevant neighborhood parameters and integrative analysis in order to optimize the neighborhoods energy management [CityGML, 2018]. In order to be able to perform large scale urban simulations like photovoltaic potential analysis, energy demand forecasts etc., it is necessary for city energy planners to link an urban three-dimensional dataset to the specific required simulation tool. This input data does not only need to represent the geometrical aspect of urban features, but also semantic information to serve simulation tools appropriately. CityGML is a useful tool for such approaches. Nevertheless, the data always need to be managed, processed and transformed in order to be used for simulations. A management framework for urban 3D data is necessary providing data on a city scale not only to one specific simulation tool, but to different tools for different scenarios.

Spatial data of developed models use vector format by which means performance of basic mathematical operations are enabled. Such quality of spatial data offers to the user's possibility to perform modeling of radio propagation conditions, "line of sight" measurements and measurements of basic urban geometry on a macro level. As a relatively new solution for fast, accurate and semi-automated three-dimensional spatial data acquisition today is offered by LiDAR (Light Detection And Ranging) technology. This technology is based on "radar like" scanning of targets, just using wavelength of light (laser). LiDAR device transmits, receive, detect and process laser waves reflected from the targets. This technology is today widely used in almost all areas of science and industry from meteorology, detection of movements, geology, physics, military to land survey. This technology provides significant situational awareness, accurate 3D representation of the scene including measurements, real-time imaging through obscuration including brownout, nighttime or otherwise hostile conditions in single 3D images or streaming videos.

Only a minority of energy management systems today extend their operation to cover multiple users and buildings. This approach demands a new set of requirements, related to multiuser access, privacy and security.

The basic purpose for the energy management systems is to monitor the energy consumption, letting the user know the historic and current energy consumption. The present most used systems include control for heating, cooling, ventilation, lightning, shadowing and electricity supply. The trend is going towards intelligent and agile systems, which have online accessibility and detailed graphical user interfaces.

Furthermore, predictive algorithms are important. More advanced functions would be to correlate consumption with user activity and outside weather conditions, optimize RES and storage systems, identify sources of high energy losses and communicate with other neighbor systems. With this extended intelligence in combination with smart meters energy management systems will be able to build up an intelligent energy management at city/districts level. Intelligent energy management systems on house level are the key technologies to achieve for example net-zero energy buildings. On the district level energy management systems are the key to enable smart micro grids and carbon-reduced cities and offer a smart platform to integrate e-mobility.

One of the main advantages of intelligent energy management systems is the use of predicted values, for weather, RES generation and consumption. These are the key data on which the predictive control algorithm and optimization strategies are based and are not satisfactorily developed yet. Investigations are required to elaborate stochastically methods for forecasting demand and production. Sophisticated forecasting enables predictive control algorithms which further lead to an optimized utilization of energy. The development of the relevant strategies should take into account an optimal utilization between RES, energy demand and eventually use of thermal electrical storages on house level under the requirements of the infrastructure on the district level. The requirements can be given by the capacitance of the infrastructure, energy efficiency issues, electrical network restrictions caused by stability and power quality aspects and reliability issues. The challenge and research effort is caused by the use of different kinds of energy (heat, electricity, gas…). Therefore, investigations concerning their synergy and divergence have to be considered as well. The design and the algorithm used in such an intelligent demand side management should handle plug and play comfort, should be adaptable to any systems structure with multiple energy resources and users and enables variable tariffs and pricing forecasts.

The integration of neighborhood energy management systems as trustworthy components in a general energy supply infrastructure, such as a smart grid for electricity, raises a number of important problems relating to security and privacy. First of all, the neighborhood energy management system must operate as a secure unit. This requires that all of the neighborhood's units operate securely and that these units interact in a secure way. Second, the microgrid must interact with the existing utilities and their regulators in a secure way. This means that the neighborhood energy management system must conform to the policies specified by the utility and the regulator, but, at the same time, the neighborhood energy system must not grant more privileges to the utility company than necessary and it must not compromise the privacy of the subscribers within the neighborhood by exposing their individual usage profiles to the utility company. This requires that the neighborhood energy management system is able to authenticate agents of the utility companies and to assign appropriate privileges to these agents, e.g., to allow the agent to perform certain operations based on his/her role in the global energy management process. In order to achieve this goal, it is necessary to develop a general security model for neighborhood energy management systems that support several levels of privileges that may be granted by multiple authorities.

3.2.2 Electricity Markets and Urban Environment

As we move on to the era of digitalization, blockchain as a technology occupies more and more space. The self-consumption matter and the use of blockchain for the intra-cities distribution networks are gaining more and more attention. So far, a centralized energy operation was the only way to run a grid. Everyone was attached to that system, either one-way or bi-directionally.

Via Blockchain and thanks to the new features that this new technology brings along, producers and consumers of energy are close to each other, and can sell and purchase energy without going through a private or state-owned company that plays the "regulator" role in some way. This Peer-to-Peer (P2P) approach is revolutionary, especially for countries where the electricity price is climbing rapidly. For instance, since few years ago, where Germany was using more Nuclear Power to produce its electricity, the electricity price was 25-50% lower. But since the country has started installing large amounts of Renewable Energy, it has officially surpassed Denmark as the most expensive country in regards to electricity prices (Fig. 3.2) [Statista, 2017].

The electricity sector is undergoing huge transformations, including a decentralized network, innovative electricity usage ways such as storage, even deep learning and IoT in coordinating electricity demand and production. But can the blockchain technology support urban areas in tackling all these hurricane-like changes in the energy sector?

The answer is Yes! Based on advancements on ICT and forecasting tools, blockchain is the way forward.

The European Union has recognized the need to move towards a low-carbon economy and has set specific and ambitious energy and climate policy-related targets for all member states. In order to achieve these targets, the EU has laid out specific technology-roadmaps that will lead to the integration of low carbon energy technologies such as solar Concentrated Solar Power (CSP) plants, Concentrated Photovoltaic (CPV), and Wind farms into the energy economy. Towards this perspective, the EU, through its Strategic Energy Technologies Information Systems (SETIS) initiative, has published the 2011 update to the Technology Map. This is the main source of reference on the current state of knowledge of low-carbon technologies in Europe. Under the SETIS initiative, the Solar Europe Industry Initiative (SEII) aims to establish a specific monitoring and knowledge sharing mechanism that includes innovative new solar forecasting algorithms and improvements within cities.

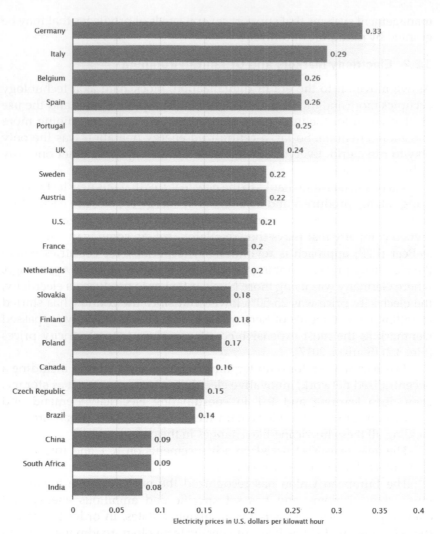

Fig. 3.2. Global electricity prices by select countries in 2017
(in U.S. dollars per kWh)

In this context, Remote Sensing and IT sectors can have a significant contribution towards the development of an efficient energy market, especially in the urban environment. They provide innovative methods to access the needed data and derive and/or exploit information products for the benefit of the energy investors and the consumers (society) respectively. Indeed as observational capabilities, and data acquisition and modeling technology moves forward, innovative research and commercial applications are further developed. Nowadays these applications take advantage of using the freely available and open access to large volumes of geo-spatial data sets provided from the operation of dedicated space based and *in-situ* environmental monitoring networks, and data portals. Moreover, the employment of advanced data modeling in combination with open data mining, data retrieval and processing tools, leads to the production of meaningful value chains for deriving reliable, accurate, stable and completely new evidences relevant to the environment, and the status of its dynamically changing parameters [Koroneos et al., 2012]. Temperature and moisture profiles, precipitation and surface properties, wind and solar energy fields, and essential atmospheric and climate variables in the urban environment can be offered for the production of advanced marketwise product solutions, creating new cross-sector business opportunities to European industries and increasing the overall competitiveness of the coupled European electricity markets sector.

3.2.3 Remote Sensing Based Applications and Forecasting

The main focus of Remote Sensing (RS) and urban focused companies is to develop advanced short-term forecasting systems ranging from every 15 minutes up to daily and/or weekly basis depending on the user's demands, taking advantage of the near real time acquisition of satellite images, and the derivation of satellite based value added products, based on various satellite missions/sensors such as MSG-SEVIRI, EOS (Aqua, and Terra)-MODIS, NOAA/AVHRR, MetOp, FY, SUOMI NPP & future JPSS, and Sentinels missions. Solar irradiation, wind speeds (onshore and offshore) in different topographies, modeling of dust dispersion phenomena, and assessment of cloud cover figures etc are taken into account on achieving accurate forecasting and in providing reliable data for simulating cities' environment.

The performance of existing forecasting tools has been analyzed in many studies, using e.g. performance indicators such as RMSE (Root Mean Square Error) and MAE (Mean Average Error). Such indicators are useful for normal operations, but much less relevant for special forecasting events. Existing forecasting tools and power planning tools are already widely used together with open RS based climate and atmospheric parameters retrieval tools, either to correlate and validate results from the current methodology where it exists, or to create completely new information products in order to ensure best wind and solar power integration so that the uncertainties can be taken into account in the wind farm and PV park control in the urban and suburban environment. This eventually leads to better models and methods for prediction of heat load and electricity generation and demand. These models are aimed often for larger groups of consumers, however are already tested on single consumer-behavior and could be extended in other areas.

As forecasting is of key importance for the success of policies that will include solar and wind power as an important contribution to cities decentralized systems, various approaches for predicting the solar irradiance and wind speeds using numerical weather models and observations, fail to achieve low uncertainty products, especially in cases of special (e.g. disaster) events. For example, for solar direct normal irradiance forecasts the statistical level of uncertainty lies within 15% when EU target towards 2020 is less than 5%. The current energy modeling approaches involve both systematic and stochastic errors. Therefore, statistical post-processing techniques are required to improve further the accuracy of power prediction. Recent efforts towards that direction consider of critical importance the integration of high-quality satellite remote sensing data together with adequate evidence provided from open geo-spatial data sources and up-to-date modeling tools. Satellite remote sensing and higher level imagery products are used to apply total sky imagery methods to retrieve cloud and aerosol information with a frequency of 15 minutes. Such products are combined with solar radiative transfer, receiver orientation models and IT tools (e.g. artificial neural networks, and rule based tools using existing knowledge from open data), in order to improve the prediction of solar energy. The current main focus of businesses now is to develop advanced and fully commercialized solar and wind energy now-casting (up to 60 minutes) and short-term forecasting (up to 4 hours) schemes taking advantage of near real time satellite images (e.g. MSG related cloud and aerosol information). Renewable energy for different components (total radiation for PV systems and direct normal for consecrated power plants), different topographies and receiver orientations, cloud motion, humidity, dust dispersion phenomena [Su et al., 2017; Xydis, 2012b; Baskut et al., 2010] taking into account on achieving accurate forecasting and providing reliable data could assist in increasing wind and solar integration rates especially within cities where most significant loads exist.

A crucial aspect to the accurate prognosis of the wind power relates to tailored and dynamic information about the conditional power production from the wind turbines and wind farms. Forecasts of wind power production are issued with different spatial resolutions i.e. turbine, park, groups of farms. These forecasts are optimized for the very short-term with varying temporal resolutions, from 15 minutes to an hour lead times. In parallel, the full description of the stochastic characteristics of wind power generation is provided, consisting of the predictive densities of power generation and of their spatial and temporal correlation via a number of parametric and nonparametric approaches.

These forecasts may be directly used as an input to the various control capabilities and to the Transmission System Operators (TSOs) to be utilized in the day-ahead and real-time markets and in Model Predictive Control (MPC) frameworks.

Modern integrated Energy Services Provider (ESPs) create a consistent system for informing on energy load forecasting and assist in market risk management, nowadays. Investors of solar and wind plants are able to actively participate in the various levels of electricity markets (day-ahead, intra-hour or real time) and try to maximize their revenue by allocating resources efficiently. Soon, they will be allowed to place bids into the market or act based on the more accurate price signal produced based on open data derived from modeling and RS solutions, offered by the system operator.

Solar and wind power-based electricity production systems can benefit greatly from accurate knowledge, predictions and data distribution, which will enable an environmentally friendly and efficient coupling between energy supply and demand together with energy supply from non-renewable systems.

The provision of high certainty energy forecast data together with the corresponding temporal and spatial estimates will have an impact on regulatory frameworks applied in each EU member state.

End-users, on the other hand, shall be able soon – based on remote sensing based demand response programs – to manage/organize their consumption based on the now casting and forecasting price signal. It is expected that both categories (investors and consumers) will have the same effectiveness in a regular non-gaming market. It will also lay the ground for new businesses (and more jobs), that will utilize the developed open data to offer services developing powerful tools that will transform real–time energy data into engaging information that shall drive energy efficiency within cities in the future.

3.3 Concluding Remarks

The need for accurate solutions for the end-users is dragging industrial innovation to its limits. Improving the performance of smart cities via a holistic system operation, which shall lead to increased efficiencies of the system and reduced operational costs, goes beyond optimization of electricity markets. The basic aim is to assist city planners and decision makers achieving new green investments and monitor smart cities operation via Key Performance Indicators (KPIs) and other assessment criteria.

The approach behind any urban based project is to couple information derived via extensive AI applications developed, via remote sensing and any other wasted information to investigate the effect of flexible loads (in abundance in cities) into every Power System, in general. Since most of developed countries continue to introduce high shares of renewables in their grids interconnected at the same time to several markets – a turn in this direction can become a paradigm for sustainable cities future urban design across EU.

REFERENCES

100% Renewable Energy is reality today, 2018. Available from: http://www. go100re.net/ (accessed on 06-Aug-2018).

Baskut, O., Ozgener, O. and Ozgener, L. (2010). Effects of meteorological variables on exergetic efficiency of wind turbine power plants. Renewable and Sustainable Energy Reviews 14(9): 3237–3241.

Best, R.E., Flager, F. and Lepech, M.D. (2015). Modeling and optimization of building mix and energy supply technology for urban districts. Applied Energy 159: 161–177.

Chalvatzis, K.J. and Rubel, K. (2015). Electricity portfolio innovation for energy security: The case of carbon constrained China. Technological Forecasting and Social Change 100: 267–276.

CityGML (2018), Available from: http://www.opengeospatial.org/standards/ citygml.

Döllner, J. and Hagedorn, B. (2007). Integrating Urban GIS, CAD, and BIM Data by Service-Based Virtual 3D City-Models. Urban Data Management Symposium. Stuttgart, Germany, pp. 157–170.

ECOFYS (2013). Towards nearly zero-energy buildings – Definition of common principles under the EPBD (Final report – Executive Summary). Available from: https://ec.europa.eu/energy/sites/ener/files/documents/nzeb_executive_ summary.pdf (accessed on 28 July 2018).

Føyn, T.H.Y., Karlsson, K., Balyk, O. and Grohnheit, P.E. (2011). A global renewable energy system: A modelling exercise in ETSAP/TIAM. Applied Energy 88(2): 526–534.

Fuchs, M., Teichmann, J., Lauster, M., Remmen, P., Streblow, R. and Müller, D. (2016). Workflow automation for combined modeling of buildings and district energy systems. Energy 117: 478–484.

Hernandez-Aramburo, C.A., Green, T.C. and Mugniot, N. (2005). Fuel consumption minimization of a microgrid. IEEE Transactions on Industry Applications 41(3): 673–681.

Keramitsoglou, I., Kiranoudis, C.T., Sismanidis, P. and Zakšek, K. (2016). An online system for nowcasting satellite derived temperatures for urban areas. Remote Sensing 8(4): 306.

Korhonen, J., Honkasalo, A. and Seppälä, J. (2018). Circular economy: The concept and its limitations. Ecological Economics 143: 37–46.

Koroneos, C.J., Nanaki, E.A. and Xydis, G.A. (2012). Sustainability indicators for the use of resources – The exergy approach. Sustainability 4(8): 1867–1878.

Koroneos, C.J., Xydis, G. and Polyzakis, A. (2010). The optimal use of renewable energy sources – The case of the new international "Makedonia" airport of Thessaloniki, Greece. Renewable & Sustainable Energy Reviews, Elsevier Publishers 14(6): 1622–1628, doi:10.1016/j.rser.2010.02.007.

Liaros, S., Botsis, K. and Xydis, G. (2016). Technoeconomic evaluation of urban plant factories: The case of Basil (Ocimum basilicum). Science of the Total Environment 554–555: 218–227.

Marnay, C., Asano, H., Papathanassiou, S. and Strbac, G. (2008). Policymaking for microgrids. IEEE Power and Energy Magazine 6(3): 66–77, May-June 2008.

Mohammadi, N. and Taylor, J.E. (2018). Smart city digital twins, 2017. IEEE Symposium Series on Computational Intelligence, SSCI 2017 – Proceedings, 2018 Jan., pp. 1–5.

Piwko, R., Meibom, P., Holttinen, H., Shi, B., Miller, N., Chi, Y. and Wang, W. (2012). Penetrating insights: Lessons learned from large-scale wind power integration. IEEE Power and Energy Magazine 10(2): 44–52.

Statista (2018). Electricity prices worldwide by country 2017. Available: https:// www.statista.com/statistics/263492/electricity-prices-in-selected-countries/

Su, X., Xu, D., Yang, R. and Yue, H. (2017). Maximum Power Point Tracking Control of Wind Turbine Considering Temperature and Humidity. Diangong Jishu Xuebao/Transactions of China Electrotechnical Society 32(13): 210–218.

Tsikalakis, A.G. and Hatziargyriou, N.D. (2008). Centralized control for optimizing microgrids operation, 2008. IEEE Transactions on Energy Conversion 23(1): 241–248.

Waymo (2018). On the road. Available from: http://www.waymo.com.

Xydis, G. (2012a). Development of an integrated methodology for the energy needs of a major urban city: The Case Study of Athens, Greece. Renewable & Sustainable Energy Reviews 16(9): 6705–6716.

Xydis, G. (2012b). Effects of air psychrometrics on the exergetic efficiency of a wind farm at a coastal mountainous site – An experimental study. Energy 37(1): 632–638.

Xydis, G. (2013a). Wind energy to thermal and cold storage – A systems approach. Energy and Buildings 56: 41–47.

Xydis, G. (2013b). The Wind Chill Temperature Effect on a large-scale PV Plant – An Exergy Approach. Progress in Photovoltaics: Research and Applications 21(8): 1611–1624.

Yu, X., You, S., Jiang, Y., Zong, Y. and Cai, H. (2017). An evolving experience learned for modelling thermal dynamics of buildings from live experiments: The Flexhouse story, 2017. Energy Procedia 141: 233–239.

Economic and Environmental Assessment of the Transport Sector in Smart Cities

E. Nanaki[1] and G. Xydis[2]

[1] University of Western Macedonia, Department of Mechanical Engineering, Bakola & Sialvera, Kozani 50100, Greece
[2] Department of Business Development and Technology, Aarhus University, Birk Centerpark 15, 7400 Herning, Denmark

4.1 Introduction

Cities around the world are important to the global economy as in terms of both production and consumption; generate a large portion of the world's GDP [Cohen, 2006]. Since the industrial revolution, the growth of cities has reached unprecedented levels. Currently 55% of the world's population lives in urban areas, a proportion that is expected to increase to 68% by 2050. Projections indicate that urbanization combined with the overall growth of the world's population could add another 2.5 billion people to urban areas by 2050 [UN, 2018].

Urban development, which is associated with the increase in urban population, has resulted in creating imbalances in complex systems such as cities. These imbalances include inter alias traffic congestion and air pollution. For this reason, it is imperative for cities to take on with the process of conversion by developing strategies to meet the challenges imposed by urbanization as well by the new demands caused by climate change and the depletion of natural resources. It is therefore crucial to manage and plan a city's urban development by supporting economic growth and competitiveness, while maintaining environmental sustainability as well as social cohesion [ARUP, 2010].

As far as air pollution is concerned, it is to be noted that in 2013 road transport reached 21% of global energy consumption and 17% of global CO_2 emissions [IEA, 2015]. Over the past decades Greenhouse Gas Emissions (GHG) from road transport, have been steadily growing and will continue to do so unless road transport is decoupled from fossil fuels [EIA, 2015]. To be more specific, unless a large-scale penetration of alternative vehicle technologies and fuels takes place, the stabilization of global temperature rise to below 2°C relative to pre-industrial levels – as set out in the 2015 Paris Agreement – will be an unrealistic target. The achievement of the above mentioned target requires an integrated approach, including improved fuel efficiency and deployment of alternative vehicle technologies as well as penetration of alternative fuels in road transport.

In the case of European Union (EU), transport is the only major sector where emissions are still increasing. While CO_2 emissions in other sectors decreased between 1990 and 2014, those from transport increased by 36% during the same period [www.Ec.europa.eu, 2014]. More than two thirds of transport-related GHG emissions come from road transport, but there are also significant emissions from the aviation and maritime sectors. This requires major changes for future mobility, as outlined by EC White Paper 2011 [European Commission, 2011], resulting in a significant decarbonization of transport to reach the 60% GHG emission reduction set for transport by 2050. The decarbonization of transport is an integral part of mitigating climate change. For this reason, many reports and guidelines in transport decarbonization policy have been published [OECD/ITF, 2015a, OECD/ITF, 2015b, UN, 2016].

Nonetheless, the major challenge of the above mentioned situation is the scaling up of (new) solutions, technologies and materials, taking into consideration the complex systems dynamics of societal and technological changes within cities. In this direction, the concept of smart cities and smart energy systems addresses improving not only mobility but also buildings, power supply, technology deployment and infrastructure towards lower emissions. Especially in the case of transport systems, alternative powertrain technologies are significant factors in mitigating carbon emissions. Overall, innovative technologies can have numerous applications, not only in improving the energy efficiency of different transport modes, but also in supporting behavioral changes, mobility patterns, and governance of transport systems.

An integrated approach, as this of a smart city, which will exploit the advantages of smart energy systems and alternative vehicle technologies will provide solutions for climate change mitigation within urban areas. This approach takes into account not only the energy system but also the innovative vehicle technologies as well as the energy efficiency improvement of the transportation system. In this direction, this chapter

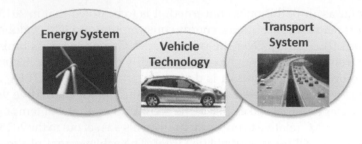

Fig. 4.1. An integrated approach to the economic and environmental assessment of transportation sector within smart cities

Color version at the end of the book

addresses key issues such as the assessment of alternative deployment approaches, within an integrated framework (Fig. 4.1).

To be more specific, the objective of this chapter is to assess innovative vehicle technologies in terms of economic and environmental sustainability for main pollution-generation sectors, through an integrated approach that will lead to higher efficiency and lower environmental impact of the energy conversion procedures and the city infrastructure integration under an effective system of system control solution. The aim of this chapter is thus to investigate the potential contribution of smart city solutions and their impact on future transport-related economic as well as environmental aspects.

4.2 The Concept of Smart City – Smart City Definition

During the past decades, many views regarding the origin of "smart city" have been expressed. In literature, 'smart', 'intelligent', 'knowledge', 'sustainable', 'digital', and 'wired' cities tend to be used interchangeably, as their definitions and principles largely overlap. However, 'smart' is the most often used label [Nam and Pardo, 2011]. The roots of the concept date back to the 1960s under what is called the 'cyber-netically planned cities'; whereas in the 1980s the term was mentioned in urban development plans, which concerned proposals for networked cities [Gabrys, 2014]. As per Dameri and Cocchia [2013] the concept was introduced in 1994. Neirotti et al. [2014] stated that the origin of the concept can be traced back to the smart growth movement in the late 1990s. Nonetheless, not until recently the concept of smart city has been adopted in city planning through the movement of smart growth [Batty et al., 2012].

Undoubtedly there are many aspects on what constitutes a smart city. These include inter alias: ecological [Lim and Liu, 2010], technological [Paroutis et al., 2014], economic [Zygiaris, 2013], organizational [Hollands, 2015] as well as societal [Deakin and Al Waer, 2012]. According to the findings of Lombardi et al. [2011] several smart cities definitions underscore the use of modern technologies and the creation of innovative transport systems, infrastructures, logistics and green and efficient energy systems. On the other hand a large body of literature emphasizes the role of human capital in developing smart cities with improved economic, social and environmental sustainability [Neirotti et al., 2014; Giffinger et al., 2007; Nam and Pardo, 2011]. The above mentioned approach considers that smart cities incorporate technological, government and social aspects, in order to enhance a smart economy, smart mobility, smart environment, smart people, smart living and smart governance [IEEE, 2014].

As a matter of fact, the concept of smart city has evolved during the past decades, from intelligent cities and it has become a "hot" topic among academics and research institutes, governments, urban planners and administration, policymakers and corporate technology firms. This evolution was mainly attributed to the challenges imposed by growing urbanization, digital revolution as well as the demands for more efficient and sustainable urban services and improved quality of life. Over the years, the concept of smart cities has incorporated variables that can provide viable solutions to a city's energy sustainability by increasing urban efficiency with regard to energy, transportation, land use, communication, economic development as well as service delivery. Indeed, a smart city represents essentially efficiency, which is based on intelligent management of urban systems using Information and Communications Technology – ICT. In this direction, diverse technologies help in achieving sustainability in smart cities [European Commission, 2012]. For this reason, smart cities and communities focus on the intersection between energy, transport and ICT. It is noted that since 2010, after the appearance of smart city projects and support by the EU, the number of publications regarding the topic has considerably increased [Jucevicius et al., 2014]. It should be noted that by 2019 the global market for smart city solutions is going to be around EUR 1.2 trillion [Transparency Market Research, 2014].

Despite the fact that the concept of smart city, is gaining publicity, the dissemination of initiatives in countries with different needs and contextual conditions hinders the identification of shared definitions and common current trends at the global scale. There is still no precise definition of the term smart city, nor a general consensus on what its describing attributes are [Caragliu et al., 2011; Neirotti et al., 2014]. In recognition of the above, the International Telecommunication Union (ITU), a United Nations specialized agency for ICTs, has listed 100 separate definitions of smart cities gathered from private companies, governments, research institutions, industry associations and NGOs activities or articles, newspapers and magazines [ITU, 2014]. Table 4.1 lists some of those definitions.

Table 4.1. Common definitions of smart cities

Authors	Definitions
Bowerman et al. (2000)	A city that monitors and integrates conditions of all of its critical infrastructures including roads, bridges, tunnels, rails, subways, airports, sea-ports, communications, water, power, even major buildings, can better optimize its resources, plan its preventive maintenance activities, and monitor security aspects while maximizing services to its citizens.
Giffinger et al. (2007)	A city well performing in a forward-looking way in [economy, people, governance, mobility, environment, and living] built on the smart combination of endowments and activities of self-decisive, independent and aware citizens.
Caragliu et al. (2009)	A city to be smart when investments in human and social capital and traditional (transport) and modern (ICT) communication infrastructure fuel sustainable economic growth and a high quality of life, with a wise management of natural resources, through participatory governance.
Eger (2009)	A particular idea of local community, one where city governments, enterprises and residents use ICTs to reinvent and reinforce the community's role in the new service economy, create jobs locally and improve the quality of community life.
Nam and Pardo (2011)	A humane city that has multiple opportunities to exploit its human potential and lead a creative life.
Zhao (2011)	Improving the quality of life in a city, including ecological, cultural, political, institutional, social, and economic components without leaving a burden on future generations.
Schaffers et al. (2012)	Smart city is referred as the safe, secure environmentally green, and efficient urban center of the future with advanced infrastructures such as sensors, electronics, and networks to stimulate sustainable economic growth and a high quality of life.
Piro et al. (2014)	A smart city is intended as an urban environment which, supported by pervasive ICT systems, is able to offer advanced and innovative services to citizens in order to improve the overall quality of their life.

(Contd.)

Table 4.1. (*Contd.*)

Authors	Definitions
Yigitcanlar (2016)	A smart city could be an ideal form to build the sustainable cities of the 21st century, in the case that a balanced and sustainable view on economic, societal, environmental and institutional development is realized.

4.2.1 Characteristics of a Smart City

Innovation combined with technologically solutions play a significant role in the concept of smart cities. A large number of studies has explored the role of innovation in increasing the competitiveness in companies, communities and cities [Fernández-Jardón et al., 2014; Mellor 2015; Kim et al., 2016]. Innovation and technology solutions that are used, in order to cover the needs of the city and its communities, will result in a smart and economically sustainable city. Nonetheless, innovation alone is not the sole ingredient of smart cities [Yigitcanlar, 2016].

Mobility, governance, environment, people and its applications as well as services (i.e. healthcare, transportation, smart education, energy) are components of cities. When these are integrated with sustainability, resilience, intelligent management of natural resources as well as with enhanced quality of life then the city is transformed to smart. Figure 4.2 illustrates the interrelationship and interdependence among the four main characteristics of a smart city: sustainability, quality of life, smartness and urbanization [Mohanty et al., 2016]. These characteristics i.e. institutional, physical, social and economic infrastructure are considered to be the main pillars of a smart city [Mohanty et al., 2016]. Sustainability aims to improve the types of activities related to services such as water, sewage, waste management, recycling, energy use and climate change mitigation, air pollution, infrastructure and governance as well as social and economic issues. Financial well-being of citizens is indicative of enhanced quality of life. Additionally, park and leisure activities, sustainable management of cultural infrastructure, as well as participation of the population in the government of the city (public referendums, etc) are crucial factors for the achievement of enhanced quality of life. In regards to urbanization, it is to be noted that this is strongly correlated with smart cities, given that factors such as the attention to urban environment, level of education, accessibility to ICT, and use of ICT in public administration influences positively the urban wealth [Caragliu et al., 2011]. Finally, the smartness of the city refers to the improvement of economic, social, and environmental standards.

Fig. 4.2. Characteristics of smart cities

In addition, the European Smart Cities initiative has established six key characteristics of smart cities [Giffinger et al., 2007]. These characteristics include: smart governance, smart people, smart mobility, smart economy, smart environment and smart living. Many studies [e.g. Batty et al., 2012; Lazaroiu, 2012] have used these characteristics, in order to develop indicators as well as frameworks and strategies.

It is obvious that there are common characteristics to many of the above mentioned features. Caragliu et al. [2011] summarized the following common features of smart cities:

- the utilization of networked infrastructure to improve economic and political efficiency and enable social, cultural, and urban development;
- an underlying emphasis on business-led urban development;
- a strong focus on the aim of achieving the social inclusion of various urban residents in public services;
- a stress on the crucial role of high-tech and creative industries in long-run urban growth;
- profound attention to the role of social and relational capital in urban development;
- social and environmental sustainability as a major strategic component of smart cities.

Based on the above mentioned, it can be stated that the characteristics of a smart city are those that can address in a "smart way" the challenges of urbanization such as waste management, air pollution, traffic congestion, health effects, resource scarcity as well as infrastructure aging. Finally, it is highlighted that the process of developing the above mentioned characteristics should take into consideration the need for leadership and organizational change, the city plan, the existence of a robust legal framework, the presence of a technological model as well as the reinforcement of business models that will ensure the effectiveness of the measures adopted.

4.3 Smart Energy Systems – Smart Mobility

The term of Smart Energy or Smart Energy Systems was first introduced in 2012 [Lund et al., 2012], in order to provide the scientific basis for a paradigm shift away from single-sector thinking into a coherent and integrated understanding of how to design and identify the most achievable and affordable strategies to implement coherent future sustainable energy systems. The term of Smart Energy has been used in many as a synonym to the term of Smart Grid [Nauman et al., 2014; Alamaniotis et al., 2016]; of Smart Grid but Broader – Cross Sector Control [Saito and Jeong, 2012]; as well as of Smart Heating [Ooka and Ikeda, 2015; Yun and Kim, 2013].

The idea of transition to smart energy systems of low carbon/zero emissions is of crucial importance, in order to mitigate the adverse effects of climate change in urban areas. The transportation sector is considered to be a key area of intervention in improving fuel quality and reducing greenhouse emissions (GHG) considering that the EU target is to reduce these emissions by 80% by 2050. Since the last decade many reports and guidelines in transport decarbonization policy have been published [OECD/ITF, 2015a, UN, 2016]. The integration of renewable energy sources to complex energy systems, such as this of transportation constitutes a great challenge towards sustainability. Facing this challenge, the need for alternative vehicle technologies as well as efficient fuel options are a necessity that cannot be disregarded. In this direction, smart mobility facilitates the achievement of the goals of a sustainable smart city by optimizing transport services, taking into account technological, societal, economic and environmental challenges.

At the EU level, several EU Polices have been introduced, in the context of clean and energy-efficient road transport vehicles, promoting directly or indirectly the electrification of road transport and smart mobility solutions. The *European Strategy for Competitive, Sustainable and Secure Energy 2020* states that the creation of market conditions which stimulate more low carbon investments into key technologies for electro-mobility, are needed [EC, 2010]. The *Roadmap on Regulation and Standards for Electric Cars* [EC, 2010a] and the respective follow-up activities aim at creating the necessary conditions for a market deployment of EVs in Europe. With Clean Power for Transport initiative [EC, 2013], the European Commission aims at the development of a comprehensive mix of alternative fuels in different transport modes. As part of this initiative the proposed Directive focuses on infrastructure and standards. It tackles one of the major obstacles for the EV market uptake – the lack of a charging infrastructure with common technical specifications by legislating a minimum number of recharging points of EVs to be installed in the Member States. The Automotive Working Group of the European Technology Platform on Smart System Integration/ERTRAC and Smart Grids [ERTRAC, 2012] has issued recommendations concerning the actual and future coverage of Research and Development topics in the field of electrification of road transport. Finally, the *European Strategy for Low-emission Mobility* has recognized three priority areas for action, in order to support Europe's transition to a low carbon circular economy. These areas include inter alias the acceleration of the deployment of low – emission alternative energy for transport as well as the removal of obstacles to the electrification of transport (i.e. improvement of electricity infrastructure) [EC, 2016a; EC, 2016b].

In this direction, zero emission transport strategies, such as the electrification of road transport is a promising option, so as to achieve the decarbonization objectives, energy security, improved urban air quality and increase energy efficiency [Department of Energy and Climate Change, 2013]. As a matter of fact, the electrification of road transport has become a major trend and EVs are becoming increasingly widespread [IEA, 2014]. Incorporating electric vehicles in cities implies a shift of the energy demand from the oil sector to the electric energy utilities, and hence the development of an adequate recharge infrastructure is needed to meet the electricity demand for electric vehicles. This further leads to upgrading the electricity source to smarter offerings.

However, there are a number of challenges for the large-scale deployment of EV both on the global and European level. These, in particular, are the high cost of the battery, lack of a standardized recharging infrastructure, relatively low range of Battery Electric Vehicles (BEV) or lack of interesting value proposition for consumers. It is noted that "range anxiety" and EV unfamiliarity may disappear as consumers are educated, but they remain strong initial obstacles to purchasing EVs [Nemry et al., 2009]. Comparing the payback periods of several advanced vehicle powertrain options versus an advanced gasoline vehicle option, it is shown that only through significant cost reductions EVs can evolve to offer an interesting value proposition to consumers [Thiel et al., 2010]. Moreover, the deployment of EVs, requires the integration of several new market actors such as vehicle producers, supply chain, charging infrastructure, providers, network operators, energy utilities and service providers with new business models or innovative vehicle to greed (V2G) solutions.

It is obvious, that the implementation of renewable energy systems in the transport sector requires various possible solutions for the needs of the future sustainable transport sector. For this reason an integrated approach exploring possible solutions to light vehicles (i.e. EVs, Hybrid Electric Vehicles, biofuels and hydrogen); heavy vehicles, aircraft and marine transport is of great significance for the creation of a low carbon transport sector. Different technologies should be combined and these should be reflected in policy strategies addressing a transformation of the total transport sector (i.e. combination of electric vehicles with the use of biofuels).

4.4 Integrated Electric Vehicle Deployment in Smart Cities

As mentioned in previous sections the electrification of road transport is of specific importance for the creation of a sustainable smart transport system, especially considering the high potential of electrified mobility for climate protection, resources management and air quality. Depending on the degree of electrification of propulsion system, the available electric drive technologies power flows for the electric vehicles are: hybrid electric, battery electric and fuel cell electric technologies.

Hybrid Electrics Vehicles (HEVs) use both an Electric Motor (EM) and an Internal Combustion Engine (ICE) to propel the vehicle. Plug-in Hybrid Electrics Vehicles (PHEVs) have a battery that can be charged off board by plugging into the grid and which enables it to travel certain kilometers solely on electricity. Battery Electrics Vehicles (BEVs) use a relatively large on-board battery to propel the vehicle. The battery provides energy for propulsion through an electric traction motor(s) as well as power for all vehicle accessory systems. Some EVs can use to drive auxiliary devices like an on board generator, which makes them have characteristics of hybrid solutions. Fuel cells are energy conversion devices set to replace combustion engines and compliment batteries in a number of applications. They convert the chemical energy contained in fuels, into electrical energy (electricity), with heat and water generated as by-products. Fuel cells continue to generate electricity for as long as a fuel is supplied, similar to traditional engines. However, unlike engines, where fuels are burnt to convert chemical energy into kinetic energy, fuel cells convert fuels directly into electricity via an electrochemical process that does not require combustion. Vehicles equipped with this technology are called Fuel Cell Electrics Vehicles (FCEVs). Electric drive technologies also, usually, incorporate other technologies, which reduce energy consumption, for example regenerative braking. This allows the electric motor to re-capture the energy expended during braking that would normally be lost. This improves energy efficiency and reduces wear on the brakes.

The electrification of road transport includes the development of Full Electric Vehicles (FEVs) specifically designed for use in the urban environment (typical daily range of 50 km), as well as Plug-in Hybrids (PHEVs) and vehicles equipped with a range extender, capable of longer trips within and between cities. Today, Internal Combustion Engines (ICE) depend heavily on fossil fuel usage, creating depletion of the finite reserve of non-renewable energy sources and creating economic and geopolitical concerns. Biofuels and natural gas are playing a role in securing fuel supply for ICEs, however just for a small fraction.

On the other hand, electricity can be produced from many different energy sources including renewable energy sources like hydro, wind solar and biomass. A large-scale rollout of electric vehicles would have an impact on the electricity system, load and foster the development of more intelligent distribution grids ('smart grids') capable of moderating such an impact. In this direction, many studies have been carried out with regards to alternative vehicle technologies, such as the electric vehicles [i.e. Nanaki et al., 2015; Nanaki and Koroneos, 2016; Ahn et al., 2011; Pina et al., 2008] including the potential integration with the electricity sector – under the smart grid concept. It is noted, that electric vehicles can facilitate the integration of more wind power by simply adjusting the time of their charge according to the renewable electricity generation, or even go a step further by both consuming and producing electricity via vehicle-to-grid (V2G) technology [Lund and Kempton, 2008]. Nonetheless, apart from the on-going technology development on the vehicle and storage side, mainly related to batteries, a wide range of technical and regulatory issues related both to the grid interconnections of EVs and the implementation of electro-mobility in public transport have to be solved prior to creating a mass market for EV. The whole system from the energy grid, through the vehicle technology and to the implementation of the transport system has to be considered and all related procedures must be developed.

Innovative mobility concepts are possible when combining electric vehicles with public transport. Nonetheless, urban mobility and electric vehicle strategies lack a holistic vision. Smart urban city planning, which is based on the concept of smart cities requires the ability to develop strategic scenarios that include the holistic mobility impact on people who are supported by the infrastructure that include also vehicles as well as the energy generation sector. Automobile industry and utility providers should work closely together with urban concept planners and city governments to understand the common mobility requirements of citizens and businesses. This would also accelerate the business service environment that is supported by applications and service stores through third party developers. The policies that promote early electric-vehicle deployment are aimed at benefits beyond near term reductions in petroleum consumption and pollutant emissions. The strategy is to speed the long-term process of conversion of the motor-vehicle fleet to alternative energy sources by exposing consumers now to electric vehicles, encouraging governments and service providers to plan for infrastructure, and encouraging the motor-vehicle industry to experiment with product design and marketing. Gaining a major market share for electric vehicles probably will require advances in technology to reduce cost and improve performance, but the premise of the early deployment efforts is that market development and technological development that proceed in parallel will lead to earlier mass adoption than if we wait for technological advances before beginning market development.

Cost-effective energy savings potential in the transportation sector from the deployment of EVs will not be fully achieved in the transportation sector as the EV market is not developing quickly enough to meet the need for increased uptake of the energy savings potential. This is because of a number of market and regulatory failures. Moreover, the adoption of eco-friendly implications and electro-mobility in urban areas are still marginalized due to economic and political barriers. In recognition of these, all stakeholders in EV play a crucial role in the delivery of a green transportation sector. A smart transport system encompasses issues such as life cycle thinking, procurement, urban planning and energy minimization. Transport systems in smart cities must take into consideration all the three pillars of sustainable development, social, economic and environmental.

4.5 Economic Assessment of Alternative Vehicle Technologies – The Case Study of EVs

The costs of alternative vehicle technologies such as electric vehicles are expected to have some differences when compared with those of conventional vehicles (Internal Combustion Engine Vehicles – ICEVs). EVs have a higher initial purchase price than CEVs due to the cost of the batteries, as well as due to uncertainties regarding the battery technology (i.e. energy capacity in relation to vehicle range, charging speed, durability, availability and environmental impacts of materials) and uncertainties regarding the vehicle. Additionally, with a larger battery and features such as regenerative braking, engine stop-start and a novel transmission system [Hutchinson et al., 2014], hybrid and electric vehicles have a higher manufacturing cost than conventional vehicles [Lave and MacLean, 2002].

The above mentioned are hindering factors to EVs acceleration and large scale deployment. It should be noted, that the adoption and deployment of EVs, is not about changing the car; it also requires changes along the energy supply chain, from the energy supply to distributed infrastructures, as well as changes in the consumers' mind sets. This represents a new paradigm, and as such, constitutes a major challenge to policy measures and instruments.

There is some literature regarding the area of ownership and life-cycle costs for vehicles; nonetheless this is a relatively new scientific area [Hagman et al., 2016]. The literature includes cost- effectiveness analysis of EVs in different European countries, where it was calculated that EVs would be cost-competitive with ICEVs in the European Union (EU) only by 2030 and coupled with costs that are 30% lower than currently expected [Seixas et al., 2015]. Cost ownership analysis of EVs in 2030 in U.S arrived at similar conclusions [Noori et al., 2015]. A comparative analysis of the lifecycle costs of electric cars to similar gasoline-powered vehicles under different scenarios of required driving range and cost of gasoline indicated that EVs with 150 km range are a technologically viable, cost competitive alternative [Werber et al., 2009]. Furthermore, the market potential of electromobility has been analyzed taking into consideration both individual priorities and barriers due to social preferences [Lieven et al., 2011].

In this direction, a comparative analysis of the Total Cost of Ownership of different EVs and ICEVs is a useful tool that can help to EVs acceleration, given that the purchase decision of vehicle customers is cost oriented. Total Cost of Ownership, takes into consideration the costs of value, taxes, insurance, maintenance, interests as well as fuel costs and allows consumers to compare all associated costs [Bubeck et al., 2016; Nanaki et al., 2015]. The following section presents the economic assessment of alternative vehicle technologies, using the TCO model, so as to derive conclusions for the future market development.

4.5.1 Total Cost of Ownership

The economic comparison between the different vehicle technologies under examination is based on the methodology of Total Cost of Ownership (TCO), which describes the costs associated over the entire lifetime of each vehicle technology. Total Cost of Ownership (TCO) refers to the sum of all costs incurred throughout the lifetime of owning or using an asset; they typically go beyond the original purchase price. TCO enables decision makers to look at asset procurement in a more strategic way (beyond the lowest bidder) and to a level playing field when choosing among competitive bids where the lowest priced bid may or may not be the least costly asset to procure. In this section, the economic comparison between power trains is based on the Total Cost of Ownership (TCO), as well as purchase price, as it describes the costs associated over their entire lifetime. All costs are "clean" of tax effects, including carbon prices. EVs are expected to have a higher purchase price than ICEs (battery related) but have a lower fuel cost (due to greater efficiency and no use of oil) and a lower maintenance cost (fewer rotating parts).

When calculating the TCO of a vehicle a number of parameters should be considered. These include:

- the purchase cost of the vehicle, including taxes and subsidies;
- the lifetime of the vehicle, or resale value after a certain number of years.
- in case of battery purchase: lifetime of the battery and, possibly, residual value.
- In case of battery lease: battery cost per kWh, or per kilometer.
- annual number of kilometers.
- fuel and/or electricity use per kilometer (in liter/km and kWh/km).
- fuel cost, including taxes.
- electricity cost, including taxes.
- maintenance cost.
- insurance cost.
- circulation tax or other taxes related to car ownership

The purchase cost and the lifetime of the vehicle determine the annual depreciation of the vehicle; whereas all the rest of the parameters, determine the annual cost of vehicle use. In addition, car owners may have to invest in a charging point at their home.

The TCO method can be applied either via a consumer-oriented approach or via a society-oriented approach [Lebeau et al., 2013]. Consumer-oriented TCO (TCO_C) models usually take into consideration the purchasing price as well as all costs related to actually receiving and using the item [Ellram and Siferd, 1998]. On the other hand, the society-oriented TCO (TCO_S) also considers the environmental costs (i.e. carbon dioxide (CO_2) emission costs) [Lebeau et al., 2013]. In the case study that will be presented, the consumer-oriented TCO is used. For this reason, in this case study all direct monetary effects crucial for the purchasing decision on new vehicles will be taken into account [Lebeau et al., 2013].

4.5.2 Case Study – TCO Model Assumptions

A comparative analysis – based on data from the Greek vehicle segment – is performed between Battery Electric Vehicles – BEV with Internal Combustion Engine-ICE vehicles (running on petrol) displaying similar characteristics, in order to provide an indication of how a typical BEV might compare to its ICE equivalent. The comparative analysis is based on real data obtained by auto-industries and on vehicles that are currently available in the market. To be more specific, the comparative analysis considers a pair of a small city car, a pair of mid-size car and a pair of a large family car. The small car segment includes vehicles of 2 doors city cars to larger 2/4 doors car, up to 1.4 L engine with a length up to 3 m; the mid-size car segment includes saloon vehicles, from 1.3 L to 2.8 L with a length up to 4.5 m. The large car segment includes executive, luxury and sport cars as well as dual and multipurpose vehicles; these come in a variety of body shapes and lengths; with the largest vehicles and engines.

Two of the vehicles are BEVs (Mitsubishi MiEV and Nissan Leaf); whereas the third (Opel Ampera REV) is an EV with range extender 18-20. The Mitsubishi MiEV is a five door, four seater all electric vehicle powered by a 47 kW permanent magnet synchronous motor. Electricity is stored in a 16 kWh Lithium-ion battery pack. The Nissan Leaf is a five door, four seat all electric vehicle powered by an 80 kW motor. Electricity is stored in a 24 kWh Lithium-ion battery pack. The Opel Ampera is an Extended-Range Electric Vehicle (E-REV), its lithium-ion battery pack powers the electric drive unit for 25 to 50 miles, which provides full vehicle speed and acceleration. For longer trips, the car's 'range-extending' engine sustains the battery. The range extender, powered by a 1.4-liter petrol-driven generator, can create electricity to power the car for a further 310 miles.

Key assumptions of the TCO vehicle model developed for this case study include:

- The average daily distance driven is considered to be 55 km, resulting to a total annual distance driven of 20,000 km.
- The average vehicle lifetime is considered 15 years or alternatively ownership of 5 years with aftersales value equal to 1/3 of the purchase price.
- Distance weighted average of ECE-15 and EUDC cycles
- The expected service life of vehicles is about 15 years
- In case the buyer applies for a loan for the purchase of the vehicle the interest rate is considered to be 8% and the payment period is set to 5 years.
- The purchase prices of the new technology vehicles take into consideration the current exemption from registration fees as well from luxury taxes.
- The fuel price cost (gasoline) is equal to €1.60 per liter and the cost of electric energy is considered to be equal to €0.12 per kWh. These conservative estimates have been utilized to simplify the economic calculations in the study. The energy prices may fluctuate significantly over the next 35 years and if anything, higher oil prices will likely lead to increased EV purchase.
- The annual running and maintenance costs of all vehicles under examination are considered to be steady during the lifetime of the vehicle.

- The cost of battery replacement during the vehicle's lifetime of 15 years has not been taken into account.
- Gasoline's combustion rate is equal to 2.4 kg CO_2 per liter. This rate is going to be used for the calculation of the CO_2 exhaust emissions.
- The average current factor of the Greek electricity mix is equal to 0.8336 kg CO_2 per kWh; whereas the combined electric energy network performance rate, charging station and charging- discharging cycle is equal to 0.88.

It should be highlighted that TCO can be influenced not only by the purchase cost of the vehicle but amongst others also by maintenance, replacement and repair costs, reliability, insurance premiums, taxes, incentives, the price for and amount of the energy that is needed for using the vehicle. The presented case study exclusively looks at purchase costs and electricity/fuel costs as average European price to the end-user. Therefore, apart from taxes and incentives, many of the above listed additional factors that could influence TCO, would most probably play a significant role against the BEV and PHEV in the beginning. For example, the higher vehicle component costs in the BEV and PHEV could lead to higher replacement costs and these again adversely influence insurance premiums. Nonetheless, through continuous improvement and learning effects these disadvantages versus the conventional vehicles would presumably reduce over time.

4.5.3 Comparative Economic Analysis of Alternative Vehicle Models

Figure 4.3 illustrates that the TCO of EVs is higher compared to the TCO of conventional vehicles. The above mentioned points out the cost-effective energy savings potential in the Greek transportation sector from the deployment of EVs will not be fully achieved in the transportation sector as the EV market has not developed quickly enough to meet the need for increased uptake of the energy savings potential. This is attributed mainly to a number of market and regulatory failures (insufficient subsidies etc.). Additionally, the adoption of eco-friendly implications and electro-mobility in urban areas are still marginalized due to economic and political barriers. Despite improvements in fuel economy, the capacity of ICEs to reduce CO_2 is significantly less than that of BEVs and FCEVs, which can achieve close to zero CO_2 emissions (wheel-to-wheel). As the range of BEVs is limited for medium sized cars, they could be ideally suited to smaller cars and shorter journeys.

Nonetheless, the lifecycle economics of EVs are expected to improve with better battery technology and mass production. The results of the TCO analysis as illustrated in Fig. 4.3 are in favor of conventional vehicles. To be more specific, the price range is between €6,307 and € 13,357 and only in the case of Nissan Leaf, the cost difference is up to €1,956, as the purchase price of Nissan Note is relatively high for a midsized vehicle. Furthermore, it is indicated at what level of purchase subsidies (or tax differentiation) the TCO of the various EVs will be equal to that of the comparable ICEs: for BVEs, about 25-35% of the purchase price would be needed; E-REVs would need about 40% of the catalog purchase price. Currently, as the findings of Fig. 4.3 reveal, BEVs and E-REVs cost more than traditional ICEVs. However, the fuel and operating costs of EVs are much lower than ICEVs, due to the high efficiency of an electric motor. Thiel et al. [2010] estimated that the payback time for a BEV, in comparison to a cheaper ICEV, would be 20 years, but should drop to less than five years by 2030.

The results of the TCO indicate that EVs are more expensive than ICE alternatives. The price difference is more likely to reduce with time as EV technology improves and ICE cost rises. Based on the above, the EV market share would depend strongly on the TCO, as sales will only increase significantly once the TCO is comparable to that of ICEs. Nonetheless, the cost of the EVs is expected to be reduced once sales volumes are increased. This is due to both economy of scale and the learning curve that is being followed. This may lead to a potential stalemate – quite a common situation for any new technology – which may be resolved by government policies. Financial policies may (temporarily) reduce the TCO of the EVs, to ensure a market share increase. Over time, the financial incentive may then be reduced as the cost of the new technology reduces. Alternatively, regulation may demand from the market to produce and sell an increasing number of the new vehicle types. This can also be expected to lead to cost reductions needed in the longer term.

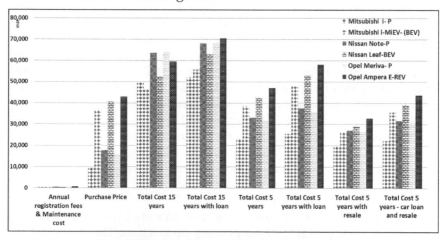

Fig. 4.3. Results of the comparative TCO analysis

4.6 Environmental Assessment of Alternative Vehicle Technologies

As already mentioned, the transport sector is responsible for almost a quarter of global energy related Greenhouse Gas Emissions – GHG. Road transportation covers more than 70%, followed by marine with a percentage of 15%, and aviation with a percentage of 10%. Urban areas, such as cities are significantly affected by the majority of vehicles that circulate daily. Currently the transportation sector is dominated by Internal Combustion Engines – ICEs. ICE engines have been widely used as engineering equipment, automobile and shipping power equipment due to their excellent drivability and economy. At the same time, ICE engines are major contributors of various types of air pollutants such as carbon dioxide (CO_2), carbon monoxide (CO), Nitrogen Oxides (NO_x), Particulate Matter (PM) and other harmful compounds.

It is highlighted, that the frequency of trips and distances, the mode of transport used as well as the technologies used by each mode significantly affect the environmental impacts of the transport sector. In this direction, several environmental policies have an objective to make a shift towards a low carbon transport system by switching the type of fuel used in vehicles to reduce emissions, promoting the use of alternative fuels such as biofuels, natural gas and electricity. In this direction, the fossil fuels used in the transport sector are expected to be replaced by alternative fuels produced from renewable energy sources, such as biofuels, biomethane, biomethanol, synthetic gasoline, and hydrogen [Meibom et al., 2013]. In view of the above, Alternative Fuel Vehicles (AFV) can tackle the challenges imposed by sustainable mobility demands over the next decades. It is to be noted, that up to date, despite the technological developments, only a small fraction of AFV's potential is tapped [Wang et al., 2015]. Technological barriers in terms of weight, volume, manufacturability, reliability, safety and durability should be tackled, in order to achieve the expected performance and efficiency [Hafez and Bhattacharya, 2017]. In 2014, around 320,000 AFVs were sold globally, and the cumulative sales amounted to 740,000 by the end of 2014. Currently, the AFV country-level market shares show a dominance in the USA, which accounts for 46% of the global AFVs sold, followed by Japan (27%), the EU (18%) and China (8%).

Life Cycle Analysis (LCA) is a useful tool that can help decision makers to achieve a holistic insight into the entire system associated with AFVs. In addition, LCA allows the comparison of materials, products or processes with different resource use and emission pathways. LCA has been used in comparative GHG emissions analysis between BEVs and ICEVs [Lombardi et al., 2017; Onat et al., 2016]. Huo et al. (2015) assessed the fuel-cycle emissions of GHGs and air pollutants (NO_x, SO_2, PM_{10}, and PM2.5) of electric vehicles in China and the United States. Comparative life cycle energy analysis between BEVs and ICEVs in China indicated that BEVs have the potential to decrease GHG emissions [Qiao et al., 2017]. In addition, LCA allows quantifying the environmental and health-related benefits and risks that may arise from shifting away from ICEVs to AFVs such as EVs [Helmers and Weiss, 2017].

In order to enable the energy transition to a low carbon transport system it is imperative to design energy and transport policies with the participation of all stakeholders, taking into account the interactions with the economy, society, environment and policy. Policy makers should design a feasible strategic plan, aiming to preserve sustainability without negatively affecting the next generations. In this direction, this section is going to assess the environmental sustainability of Alternative Fuel Vehicles (AFVs), using Life Cycle Analysis. The technologies considered are Hybrid Electric Vehicles (HEVs), Plug-in Hybrid Electric Vehicles (PHEVs), Electric Vehicles (EVs) and Fuel Cell Vehicles (FCVs). These are promising technologies that could replace a significant portion of the global fleet's petroleum consumption and mitigate emissions and security issues related to oil extraction, importation and combustion [Nanaki and Koroneos, 2016].

4.6.1 Life Cycle Analysis

LCA tools have become more efficient and robust for the process of identification and quantification of potential environmental burdens and impacts of a product, process or an activity [Jeswani et al., 2010]. During the last decades, this methodology has been applied in different sectors including the field of agricultural, engineering and material sciences, thereby becoming an invaluable decision-support tool for manufacturers, policy-makers and other stakeholders. As per International Organization for Standardization (ISO) and the Society for Environmental Toxicology and Chemistry (SETAC), the LCA methodology consists of the following stages [ISO 14040–ISO 14044, 2006]: 1) Goal and scope definition; 2) Life Cycle Inventory (LCI) analysis; 3) Life Cycle Impact Assessment (LCIA); and finally, Life Cycle Interpretation.

Energy efficiency and pollutant emissions are the main drivers in comparative analysis of different vehicle technologies. The LCA method through a "cradle-to-grave" approach evaluates different vehicle technologies through their life cycle and identifies the related environmental burdens taking into account the construction, operation, and decommissioning phases [Torchio and Santarelli, 2010]. The identification and quantification of potential environmental burdens related to the raw materials used, the energy used, and the emissions released to the environment are the main characteristics of an LCA. The analysis of both energy efficiency and GHG emissions can provide useful conclusions in terms of primary sources exploitation, and reduction of GHG emission in the overall fuel cycle. In automotive life cycle analysis (Fig. 4.4) two stages are discerned: the vehicle life cycle and fuel life cycle. Vehicle life cycle refers to material production, vehicle assembly, distribution, and disposal; whereas the fuel life cycle (also known as well-to-wheels analysis) can be divided into two stages: the well-to-tank (energy consumption and emissions to extract raw materials, to transport them, to produce the desired fuel, to distribute the fuel to consumers, and so on); and tank-to wheels (energy consumption and emissions caused by using the fuel by vehicle) [Torchio and Santarelli, 2010].

Modeling of life cycle energy and GHG emsisions is made using GREET [https://greet.es.anl.gov/]. The open-source software consists of multidimensional worksheets that have been developed to address analytical issues associated with energy and emission impacts of advanced and new transportation fuels. In regards to disposal and recycling, the GREET model takes into account the energy use and emissions during recycling of scrap materials back into its original and reusable form. For comparative purposes, the analysis of the AFVs is referred to the same functional unit. The chosen functional unit for the study was the cradle-to-grave (excluding end-of-life management) life cycle of a compact passenger vehicle, including an assumed 150,000 km use phase.

4.6.2 Alternative Fuel Vehicles Specifications

The selected reference vehicle is a conventional Volkswagen Golf compact passenger car [www.volkwagen.com] running on petrol. The reference vehicle uses only an internal combustion engine to drive the wheels through a mechanical transmission. Nissan Leaf [www.nissanusa.com] is chosen as EV representative. Toyota Prius [www.toyota.com], is selected as a representative HEV. It is noted, that Toyota Prius employs a power-split propulsion system. The Toyota Prius Plug-in Hybrid selected as HPHEV, is similar to the conventional HEV (i.e. Toytoa Prius), but it differs from the HEV due to its higher battery energy capacity, the utilization of an external power source to recharge the batteries, and the

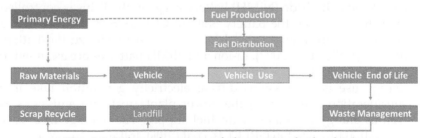

Fig. 4.4. Life cycle stages of alternative fuel vehicles
Color version at the end of the book

different battery management strategy. Toyota Prius Plug – in Hybrid can be recharged using an electric connection [www.toyota.com]. It uses a smaller battery pack. Even in charge depleting mode, the vehicle turns the engine on whenever needed to assist the motor (e.g., full acceleration) and/or charge the batteries because the propulsion motor is not sized for peak power requests. The configuration of Honda FCV Clarity, selected as FCV [www.automobiles.honda.com] is similar to a series HEV with the difference that the primary energy source is a fuel cell rather than an engine-generator. Hydrogen, stored in high pressure tanks, is the fuel for fuel cells. Hydrogen is produced via steam reforming of natural gas at petroleum-refining facilities and then distributed to fuel stations for vehicle use.

Table 4.2 summarizes the specifications of the vehicle technologies selected for this case study. It is assumed that the vehicles are built on a similar platform and therefore, the energy use and emissions of the vehicles are comparable throughout their lifetimes. The following assumptions have been made:

- The chosen functional unit is the cradle-to-grave (excluding end-of-life management) life cycle of a compact passenger vehicle, including an assumed 150,000 km use phase.
- The vehicles are assumed to have a lifetime of 150,000 km
- All vehicle manufacturers use the same processes to build vehicles. The GREET model is employed, in order to estimate energy requirements and emissions from the manufacturing process.
- For the ICEV, the use phase fuel consumption is assumed to be equal to the officially claimed in the figure for the 1.4 L Volkswagen Golf, i.e., 6.2 L (petrol) per 100 km under Federal Test Procedure -75 (FTP -75) conditions. The energy consumption of the HEV, PEV, EV and FCV is assumed to be 15 kWh (electricity) per 100 km under FTP -75 conditions [Nissan, 2017].
- The hybrid system, including the high-voltage battery pack, is warranted for 8 years/160,000 km [Nissan Leaf Battery, 2017].
- Although Li-ion batteries have more charge/discharge cycles, the nickel–metal hydride (NiMH) battery usage in the Prius is controlled in such a way that the battery State of Charge (SOC) is forced to a target value throughout the driving cycle to maximize the battery lifetime [Kelly et al., 2002]. Li-ion and NiMH batteries are assumed to be replaced at the same time.
- Energy use and emissions during electricity generation take into consideration emissions at the power plant, upstream emissions as well as energy associated with fuel supply. The average electricity generation mix used in GREET is: coal (51%), followed by natural gas (19%), nuclear (18%), and others (12%) [Wang et al., 2007].

Table 4.2. Technical Specifications of Reference and Alternative Fuel Vehicles

	UNIT	PETROL	HEV	PEV	EV	FCV
Model		Volkswagen Golf	Toyota Prius	Toyota Prius Plug In	Nissan Leaf	Honda Clarity
Dimensions (L)	m	4.21	4.46	4.48	4.47	4.83
H*W	m	1.48*2.0	1.49*1.75	1.49*1.75	1.55*1.77	1.47*1.85
Weight	kg	1342	1381	1437	1528	1626
Engine Power	kW	100	73	73	-	100[a]
Motor Power	kW	-	60	60	80	100
Total Power	kW	100	100	100	80	100
Battery Type[b]		-	Ni-MH	Li-ion	Li-ion	Li-ion
Battery Weight	kW	-	50	150	300	23
Battery Energy	kWh	-	1.3	4.4	24	2
All Electric Range	km	-	-	24	117	-
Price[c]	eur	19,500	32,000	37,000	35,200	59,385

a 100 kW hydrogen fuel cell stack
b Refers to high-voltage battery used to drive electric motors
c Basic model price for the year 2015 and 2016

4.6.3 Comparative Life Cycle Analysis of Alternative Vehicle Technologies

Figures 4.5 and 4.6 illustrate the lifetime energy consumption and GHG emissions of ICE and AFVs. The fuel production cycle includes fuel and feedstock stages. The results of the LCA indicate that the ICEV has the highest energy use and GHG emissions per km, most of which are produced during the vehicle operation. The HEV has the lowest energy use and GHG emissions per km. HEV improves the fuel economy at the vehicle operation phase by eliminating idling losses, downsizing engine displacement, operating the engine at nearly optimal conditions, and recovering energy with regenerative braking [Huang and Zhang, 2011].

Regarding vehicles' materials, it is noticed that the production of the fuel-efficient vehicles is a more energy-intensive process due to the complex electrical systems. Figures 4.5 and 4.6 indicate that larger battery sizes and more powerful electronics result in more energy use and GHG emissions during the stage of vehicle production.

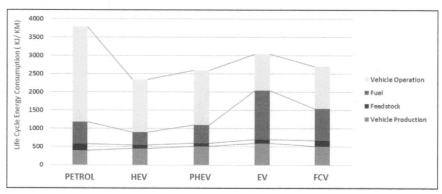

Fig. 4.5. Life cycle energy consumption of alternative vehicle technologies

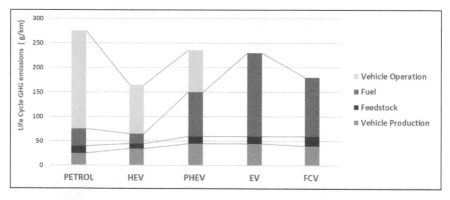

Fig. 4.6. Life cycle GHG emissions of alternative vehicle technologies

It is noted that the impact of EVs is indeed zero for urban air pollution. Nonetheless, if the emissions from the production of electricity in the power stations and from the manfacture of the vehicles is taken into consideration, then the total amount of GHG emissions of the EV is only less than the emissions of the CV, but the emissions created during the fuel usage stage are much more than those by different fuel pathways. It is highlighted that in a country where most of the electricity is produced by burning coal there would be only minor GHG emission benefit from the EV introduction [Nanaki and Koroneos, 2013]. The largest reduction is associated with the use of renewable energies with the lowest values for EVs achieved e.g. in the emerging "carbon free communities", where the electricity is produced entirely by wind, water, photovoltaic, geothermal energy, biomass or animal waste. However, in a vision where most of new power installations will be renewable technologies, the EV is considered a way towards a radical reduction of greenhouse gas emissions. Deployment of electric vehicles may even help to extend the use of renewable energy if it is targeted at captive fleets in areas close to an abundant supply of stochastic renewable electricity. With the mass deployment of EVs the securing of marginal energy production from sources with low GHG emissions will become important for the GHG balance of EVs.

It is estimated that approximately 48, 28, 15 and 9% of energy and emissions are coming from the stage of battery production during the manufacture of EVs, PHEVs, HEVs and FCVs respectively. The results also indicate that the total energy usage of the FCV is higher than that of both the HEV and PHEV, while the GHG emissions are only higher than those of the HEV. Fuel cells have high energy conversion efficiencies from hydrogen to electricity of up to 83%. In automotive applications, however, a fuel cell needs a peripheral system, which lessens the net efficiency to approximately 52% [Thomas, 2009].

4.7 Concluding Remarks

Smart cities and emerging technologies are key factors to climate change mitigation and to the transformation of the transport energy system. Overall, innovative technologies can have numerous applications, not only in improving the energy efficiency of different transport modes, but also in supporting behavioral changes, mobility patterns, and governance of transport systems. Future transport and mobility services cannot be considered as stand-alone sectoral solutions. Given the significant urbanization not only in Europe, but in other regions, these will need to be embedded in wider smart and sustainable city strategies aimed at increasing urban resource efficiency and decarbonization including the development and implementation of new and clean technologies.

Alternative Fuel Vehicles – AFVs have the potential to provide an integrated strategic transition plan to the transition to zero air pollution/ low carbon societies. Taken into consideration the fact that the goal is the widespread deployment of AFVs in urban areas and the uptake of cleaner fuels, in order to reduce air emissions, it is critical to identify and evaluate the barriers to their adoption and to develop new strategies. In this context, emphasis should be given on research and innovation on delivering solutions for the challenges that the deployment of AFVs in urban areas face. The creation of an integrated methodological framework, which incorporates the concepts of "smart cities" and "smart grids" is of great significance. In order to have a real impact on the green economy, research in the field of AFVs should no longer focus on alternative vehicle technologies seen in isolation from the rest of the transport and energy system: a massive introduction of the technology requires the availability of smart electricity grids and intelligent vehicle charging systems tailored to customer's needs. Cost-effective energy savings potential in the transportation sector from the deployment of AFVs will not be fully achieved in the transportation sector as the market is not developing quickly enough to meet the need for increased uptake of the energy savings potential. This is because of a number of market and regulatory failures. Moreover, the adoption of eco-friendly implications and electro-mobility in urban areas are still marginalized due to economic and political barriers.

In recognition of these, all stakeholders play a crucial role in the delivery of a green transportation sector. A sustainable transport system encompasses issues such as life cycle thinking, urban design, urban planning and organization as well as energy minimization. Sustainable urban transport systems must take into consideration all three pillars of sustainable development, social, economic and environmental. Moreover, it is clear that the declared environmental targets cannot be met without dramatically reducing the environmental impact of energy utilization in the transportation sector.

REFERENCES

Ahn, C., Li, C.T. and Peng, H. (2011). Optimal decentralized charging control algorithm for electrified vehicles connected to smart grid. Journal of Power Sources 196: 10369–10379.

Alamaniotis, M., Bargiotas, D. and Tsoukalas, L.H. (2016). Towards smart energy systems: Application of kernel machine regression for medium term electricity load forecasting. Springer plus 5.

ARUP (2010). Smart Cities. Transforming the 21st Century City via the Creative Use of Technology. Available at: https://www.arup.com/perspectives/ publications/research/section/smart-cities

Batty, M., Axhausen, K.W., Giannotti, F., Pozdnoukhov, A., Bazzani, A., Wachowicz, M., Ouzounis, G. and Portugali, Y. (2012). Smart cities of the future. The European Physical Journal 214: 481–518.

Bowerman, B., Braverman, J., Taylor, J., Todosow, H. and Wimmersperg, U. (2000). The vision of a smart city. *In*: 2nd International Life Extension Technology Workshop, Paris.

Bubeck, S., Tomascheck, J. and Fahl, U. (2016). Perspectives of electric mobility: Total cost of ownership of electric vehicles in Germany. Transport Policy 50: 63–77.

Caragliu, A., Del Bo, C. and Nijkamp, P. (2009). Smart cities in Europe. Series Research Memoranda 0048. Faculty of Economics, Business Administration and Econometrics. VU University Amsterdam.

Caragliu, A., Del Bo, C. and Nijkamp, P. (2011). Smart cities in Europe. Journal of Urban Technology 18: 65–82.

Cohen, B. (2006). Urbanization in developing countries: Current trends, future projections, and key challenges for sustainability. Technology in Society 28: 63–80.

Cummings, M.L. and Thornburg, K.M. (2011). Paying attention to the man behind the curtain. IEEE Pervasive Computing 10(1): 58–62.

Dameri, R. and Cocchia, A. (2013). Smart city and digital city: Twenty years of terminology evolution. X Conference of the Italian Chapter of AIS, ITAIS 2013, Università Commerciale Luigi Bocconi, Milan (Italy), pp. 1–8.

Deakin, M. and Al Waer, H. (2011). From intelligent to smart cities. Intelligent Buildings International 3(3): 140–152.

Department of Energy and Climate Change (March 2013). Statistical Release. <www.decc.gov.uk/assets/decc/11/stats/climate-change/4817-2011-uk-greenhouse-gas-emissions-provisional-figur.pdf>.

Eger, J.M. (2009). Smart growth, smart cities, and the crisis at the pump a worldwide phenomenon. The Journal of E-Government Policy and Regulation 32(1): 47–53.

Ellram, L.M. and Siferd, S.P. (1998). Total cost of ownership: A key concept in strategic cost management decisions. Journal of Business Logistics 19: 55–84.

Energy Information Administration – EIA (2015). Total Carbon Dioxide Emissions from the Consumption of Energy (Million Metric Tons). [Online] Available at: https://www.eia.gov/cfapps/ipdbproject/IEDIndex3.cfm?tid =90&pid=44&aid=8

ERTRAC/EPoSS/SmartGrids (2012). European Roadmap. Electrification of Road Transport. 2nd Edition.

European Commission (2010). Energy 2020. A strategy for competitive, sustainable and secure energy. COM (2010) 639 final, Brussels.

European Commission (2010a). Roadmap on regulations and standards for the electrification of cars. Available online at: https://www.ertrac.org/uploads/images/ERTRAC_ElectrificationRoadmap2017.pdf

European Commission (2011). White Paper, Roadmap to a Single European Transportation Area – Towards a competitive and resource efficient transport system, s.l.: s.n.

European Commission (2012). Communication from the Commission. Smart cities and communities – European innovation partnership. Brussels. Available online at: http://ec.europa.eu/energy/technology/initiatives/doc/ 2012_4701_smart_cities_en.pdf

European Commission (2013). Clean Power for Transport: A European alternative fuels strategy. COM(2013) 17 final Brussels.

European Commission (2014). Reducing emissions from transport. Available from: http://ec.europa.eu/clima/policies/transport/index_en.htm

European Commission (2016a). Communication from the Commission to the European Parliament, the Council, the European Economic and Social Committee and the Committee of the Regions. A European Strategy for Low-emission Mobility COM/2016/0501 final.

European Commission (2016b). Communication from the Commission to the European Parliament, the Council, the European Economic and Social Committee the Committee of the Regions and the European Investment Bank. Clean Energy for All Europeans COM(2016) 860 final.

Fernández-Jardón, C., Costa, R.V. and Dorrego, P.F. (2014). The impact of structural capital on product innovation performance: An empirical analysis. International Journal of Knowledge-Based Development 5(1): 63–79.

Gabrys, J. (2014). Programming environments: Environmentality and citizen sensing in the smart city. Environment and Planning D: Society and Space 32: 30–48, 10.1068/d16812.

Giffinger, R., Fertner, C., Kramar, R., Kalasek, N., Pichler-Milanović, E. and Meijers. (2007). Ranking of European medium-sized cities. Centre of Regional Science, Vienna UT. Available online at: http://www.smartcities.eu/download/ smart_cities_final_report.pdf

GREET: https://greet.es.anl.gov/

Hafez, O. and Bhattacharya, K. (2017). Optimal design of electric vehicle charging stations considering various energy resources. Renewable Energy 107: 576–589.

Hagman, J., Ritzén, S., Stier, J.J. and Susilo, Y. (2016). Total cost of ownership and its potential implications for battery electric vehicle diffusion. Research in Transportation Business & Management 18: 11–17.

Helmers, E. and Weiss, M. (2017). Advances and critical aspects in the life-cycle assessment of battery electric cars. Energy and Emission Control Technologies 5: 1–18.

Hollands, R.G. (2015). Critical interventions into the corporate smart city. Cambridge Journal of Regions, Economy and Society 8(1): 61–77.

Honda Clarity Fuel Cell Car Homepage. Available online: http://automobiles. honda.com/fcxclarity/

http://ec.europa.eu/enterprise/sectors/automotive/files/pagesbackground/ competitiveness/roadmap-electric-cars_en.pdf

Huang, W.-D. and Zhang, Y.-H.P. (2011). Energy efficiency analysis: Biomass-to-wheel efficiency related with biofuels production, fuel distribution, and powertrain systems. PLoS One 6: e22113.

Huo, H., Cai, H., Zhang, Q., Liu, F. and He, K. (2015). Life-cycle assessment of greenhouse gas and air emissions of electric vehicles: A comparison between China and the U.S. Atmospheric Environment 108: 107–116.

Hutchinson, T., Burgess, S. and Herrmann, G. (2014). Current hybrid-electric powertrain architectures: Applying empirical design data to life cycle assessment and whole-life cost analysis. Applied Energy 119: 314.

IEEE (The Institute of Electrical and Electronics Engineers) (2014). IEEE smart cities http://smartcities.ieee.org/about.html

International Energy Agency – IEA (2014). EV City Casebook. Available from: <http://www.cleanenergyministerial.org/Portals/2/pdfs/EVI_2014_EV-City-Casebook.pdf>

International Energy Agency Statistics – IEA (2015). CO_2 Emissions from fuel combustion: Highlights, Paris.

ISO 14040 (1997). Environmental Management—Life Cycle Assessment—Principles and Framework. Geneva: International Organization for Standardization.

ISO 14041 (1998). Environmental Management—Life Cycle Assessment—Goal and Scope Definition and Inventory Analysis. Geneva: International Organization for Standardization.

ISO 14042 (2000). Environmental Management—Life Cycle Assessment—Life Cycle Impact Assessment. Geneva: International Organization for Standardization.

ISO 14043 (2000). Environmental Management—Life Cycle Assessment—Life Cycle Interpretation. Geneva: International Organization for Standardization.

ISO 14044 (2006). Environmental Management Life Cycle Assessment Requirement and Guidelines. Geneva: International Organization for Standardization.

ITU (2014). Smart sustainable cities: An analysis of definitions by ITU-T Focus Group on Smart Sustainable Cities. from http://www.itu.int/en/ITU-T/focusgroups/ssc/Pages/default.aspx.

Jeswani, H.K., Azapagic, A., Schepelmann, P. and Ritthoff, M. (2010). Options for broadening and deepening the LCA approaches. Journal of Cleaner Production 12: 120–127.

Jucevicius, R., Pataˇsienˉe, I. and Pataˇsius, M. (2014). Digital dimension of smart city: Critical analysis. Procedia – Social and Behavioral Sciences 156: 146–150.

Kelly, K.J., Mihalic, M. and Zolot, M. (2002). Battery usage and thermal performance of the Toyota Prius and Honda Insight during chassis dynamometer testing. pp. 247–252. *In*: Proceedings of the 17th Battery Conference on Applications and Advances. Long Beach, CA, USA.

Kim, S.J., Kim, E.M., Suh, Y. and Zheng, Z. (2016). The effect of service innovation on R&D activities and government support systems: The moderating role of government support systems in Korea. Journal of Open Innovation: Technology, Market, and Complexity 2(1): 1–19.

Lave, L.B. and MacLean, H.L. (2002). An environmental-economic evaluation of hybrid electric vehicles: Toyota Prius vs. its conventional internal combustion engine Corolla. Transportation Research Part D: Transport and Environment 7: 155.

Lazaroiu, G.C. (2012). Definition methodology for the smart cities model. Energy 20(1): 326–335.

Lebeau, K., Lebeau, P., Macharis, C. and Van Mierlo, J. (2013). How expensive are electric vehicles? A total cost of ownership analysis. World Electric Vehicle Symposium and Exhibition (EVS27), ISBN 978-1-4799-3833-9, 4: 2213–2224.

Lieven, T., Mühlmeier, S., Henkel, S. and Waller, J.F. (2011). Who will buy electric cars? An empirical study in Germany. Transportation Research, Part D 16: 236–243.

Lim, C.J. and Liu, E. (2010). Smartcities and Eco-warriors. New York: Routledge.

Lombardi, L., Tribioli, L., Cozzolino, R. and Bella, G. (2017). Comparative environmental assessment of conventional, electric, hybrid, and fuel cell power trains based on LCA. The International Journal of Life Cycle Assessment 12: 1–18.

Lombardi, P., Giordano, S., Caragliu, A., Del Bo, C., Deakin, M., Nijkamp, P. and Kourtit, K. (2011). An advanced triple-helix network model for smart cities performance. Vrije Universiteit Amsterdam, Research Memorandum. Available online at: http://degree.ubvu.vu.nl/repec/vua/wpaper/pdf/20110045.pdf

Lund, H., Andersen, A.N., Østergaard, P.A., Mathiesen, B.V. and Connolly, D. (2012). From electricity smart grids to smart energy systems – A market operation based approach and understanding. Energy 42: 96–102.

Lund, H. and Kempton, W. (2008). Integration of renewable energy into the transport and electricity sectors through V2G. Energy Policy 36: 3578–3587.

Meibom, P., Hilger, K.B., Madsen, H., and Vinther, D. (2013). Energy comes together in Denmark. Power & Energy Magazine 11(5): 46–55.

Mellor, R.B. (2015). Modelling the value of external networks for knowledge realization, innovation, organizational development and efficiency in SMEs. International Journal of Knowledge-Based Development 6(1): 3–14.

Mohanty, S.P., Choppali, U. and Kougianos, E. (2016). Everything you wanted to know about smart cities: The internet of things is the backbone. IEEE Consumer Electronics Magazine 5: 60–70.

Nam, T. and Pardo, T.A. (2011). Conceptualizing smart city with dimensions of technology, people and institutions. pp. 282–291. *In*: Proceédings of the 12th Annual International Digital Government Research Conference: Digital Government Innovation in Challenging Times.

Nanaki, E.A. and Koroneos, C.J. (2016). Climate change mitigation and deployment of electric vehicles in urban areas. Journal of Renewable Energy 99: 1153–1160.

Nanaki, E.A. and Koroneos, C.J. (2013). Comparative economic and environmental analysis of conventional, hybrid and electric vehicles – The case study of Greece. Journal of Cleaner Production 53: 261–266.

Nanaki, E.A., Xydis, G.A. and Koroneos, C.J. (2015). Electric Vehicle Deployment in Urban Areas. Journal of Indoor and Built Environment 25(7): 1065–1074.

Naumann, A., Bielchev, I., Voropai, N. and Styczynski, Z. (2014). Smart grid automation using IEC 61850 and CIM standards. Control Engineering Practice 25.

Neirotti, P., De Marco, A., Cagliano, A.C., Mangano, G. and Scorrano, F. (2014). Current trends in smart city initiatives – Some stylized facts. Cities 38: 25–36.

Nemry, F., Leduc, G. and Munoz, A. (2009). Plug-in Hybrid and Battery-Electric Vehicles: State of the Research and Development and Comparative Analysis of Energy and Cost Efficiency. Luxembourg: Institute for Prospective Technological Studies. European Commission Joint Research Centre.

Nissan Leaf Battery & Performance (2017). https://www.nissan.co.uk/vehicles/new-vehicles/leaf/battery-performance.html

Nissan Leaf Electric Car Homepage. Available online: http://ww.nissanusa.com/leaf-electric-car

Noori, M., Gardner, S. and Tatari, O. (2015). Electric vehicle cost, emissions, and water footprint in the United States: Development of a regional optimization model. Energy 89: 610–625.

Onat, N.C., Kucukvar, M., Tatari, O. and Egilmez, G. (2016). Integration of system dynamics approach toward deepening and broadening the life cycle sustainability assessment framework: A case for electric vehicles. International Journal of Life Cycle Assessment 21: 1009–1034.

Ooka, R. and Ikeda, S. (2015). A review on optimization techniques for active thermal energy storage control. Energy Buildings 106.

Organization for Economic Co-operation and Development – OECD/ITF (2015a). Shifting towards Low Carbon Mobility Systems. Paris.

Organization for Economic Co-operation and Development – OECD/ITF (2015b). Policy Strategies for Vehicle Electrification. Paris.

Paroutis, S., Bennett, M. and Heracleous, L. (2014). A strategic view on smart city technology: The case of IBM Smarter Cities during a recession. Technological Forecasting and Social Change 89: 262–272.

Pina, A., Ioakimidis, C.S. and Ferrao, P. (2008). Introduction of electric vehicles in an island as a driver to increase renewable energy penetration. pp. 1108–1113. *In*: 2008 IEEE International Conference on Sustainable Energy Technologies, ICSET.

Piro, G., Cianci, I., Grieco, L.A., Boggia, G. and Camarda, P. (2014). Information centric services in smart cities. Journal of Systems and Software 88(1): 169–188.

Qiao, Q., Zhao, F., Liu, Z., Jiang, S. and Hao, H. (2017). Cradle-to-gate greenhouse gas emissions of battery electric and internal combustion engine vehicles in China. Applied Energy 204: 1399–1411.

Saito, K. and Jeong, J. (2012). Development of general purpose energy system simulator. Energy Procedia 14: 1595–1600.

Schaffers, H., Komninos, N., Pallot, M., Aguas, M., Almirall, E., Bakici, T., Barroca, J., Carter, D., Corriou, M. and Fernadez, J. (2012). Smart cities as innovation ecosystems sustained by the future Internet. Technical Report pp. 65. hal-00769635

Seixas, J., Simoes, S., Dias, L., Kanudia, A., Fortes, P. and Gargiulo, M. (2015). Assessing the cost-effectiveness of electric vehicles in European countries using integrated modeling. Energy Policy 80: 165–176.

SETAC Foundation (1991). A technical framework for life cycle assessment. Washington, D.C.: Society of Environmental Toxicology and Chemistry and SETAC Foundation for Environmental Education, Inc.

Thiel, C., Perojo, A. and Mercier, A. (2010). Cost and CO_2 aspects of future vehicle options in Europe under new energy policy scenarios. Energy Policy 38: 7142–7151.

Thomas, C.E. (2009). Fuel cell and battery electric vehicles compared. International Journal of Hydrogen Energy 34: 6005–6020.

Torchio, M.F. and Santarelli, M.G. (2010). Energy, environmental and economic comparison of different powertrain/fuel options using well-to-wheels assessment, energy and external costs – European market analysis. Energy 35: 4156–4171.

Toyota 3rd Generation Prius Homepage. Available online: http://www.toyota.com/prius-hybrid/

Toyota Prius Plug-in Hybrid Homepage. Available online: http://www.toyota.com/prius-plug-in/

Transparency Market Research (2014). Global smart cities market: Industry analysis, size, share, growth, trend and forecast 2013–2019. Transparency Market Research. Retrieved from http://www.transparencymarketresearch.com/smart-cities-market.htm

United Nations – UN (2016). Mobilizing Sustainable Transport for Decarbonization Transport for Development. NY

United Nations – UN (2018). Department of Economic and Social Affairs. https://www.un.org/development/desa/en/news/population/2018-revision-of-world-urbanization-prospects.html (accessed on October 2018).

Volkswagen webpage. Available online: www.volkswagen.com

Wang, H., Zhang, X. and Ouyang, M. (2015). Energy consumption of electric vehicles based on real-world driving patterns: A case study of Beijing. Applied Energy 157: 710–719.

Wang, M., Wu, Y. and Elgowainy, A. (2007). Operating Manual for GREET: Version 1.7. Technical Report ANL/ESD/05–3; Argonne National Laboratories: Lemont, IL, USA.

Werber, M., Fischer, M. and Schwartz, P.V. (2009). Batteries: Lower cost than gasoline? Energy Policy 37: 2465–2468.

Yigitcanlar, T. (2016). Technology and the City: Systems, Applications and Implications. New York: Routledge.

Yun, J. and Kim, J.(2013). Deployment support for sensor networks in indoor climate monitoring. International Journal of Distributed Sensor Networks 9(9), https://doi.org/10.1155/2013/875802

Zhao, J. (2011). Towards Sustainable Cities in China: Analysis and Assessment of Some Chinese Cities in 2008. Berlin: Springer.

Zygiaris, S. (2013). Smart city reference model: assisting planners to conceptualize the building of smart city innovation ecosystems. Journal of the Knowledge Economy 4: 217. https://doi.org/10.1007/s13132-012-0089-4

Transportation and Urban Environment – Exergetics

E. Nanaki[1] and G. Xydis[2]

[1] University of Western Macedonia, Department of Mechanical Engineering,
Bakola & Sialvera, Kozani 50100, Greece
[2] Department of Business Development and Technology, Aarhus University,
Birk Centerpark 15, 7400 Herning, Denmark

5.1 Introduction

The transportation sector plays a significant role in the creation of a sustainable energy future. Taking into consideration the fact that this sector has many intercorrelations with economic and social activities, it is easily deducted that this sector has a substantial impact on and, consequently, influence over almost all environmental matters. As a matter of fact, the transportation sector is one of the most challenging, when trying to achieve the goals towards the creation of a low carbon transportation system [Mathiesen et al., 2015]. The transportation sector is the second largest energy consumer sector – after the industrial-accounting for 30% of the growth in petroleum consumption between 2004 and 2030 [IEA, 2012]. It is noted that during the past decade, one third of the total final energy consumption and more than one fifth of greenhouse gas (GHG) emissions in the European Union (EU) were attributed to the consumption of fossil fuels in the transport sector [Alises and Vassalo, 2015]. As per the World Energy's Statistics [IEA, 2014], the total energy required by the global transport system worldwide rose from 23% in 1973 to 28% in 2012. In 2050, as much as 30–50% of total CO_2 emissions will come from the transportation sector [Fluglestvedt et al., 2007], compared to 22% in 2008 [Geng et al., 2013].

Cities constitute a critical part of the transport system, as more than 72% [United Nations, 2007] of Europeans live in urban areas and this percentage is expected to increase. Cities are the powerhouse of economic growth and development, since around 85% of the EU's GDP is generated in urban areas. At the same time, 40% of total CO_2 emissions and 70% of emissions of other pollutants are caused by urban traffic. Challenges like road traffic congestion, road safety, environmental impacts, urban sprawl, and increasing demand for mobility (mainly satisfied by private car ownership) are common to many European cities.

The need to solve these problems has become even more crucial to maintaining a high quality of life in a sustainable way, competitive as well as smooth mobility of people and goods. Sustainability is thus the main goal that underpins current approaches and solutions for future mobility. Sustainability should lie at the heart of all policies and strategies for a more sustainable transport system both in environmental and competitiveness terms, while also addressing social concerns. This is why the concept of sustainability goes far beyond the need to respond by managing road traffic flows and their impacts, because it should also address, for instance, the cost of mobility in relation to social exclusion, economic and social cohesion, and the demographic changes that will shape the structure of cities in the future.

In regards to the investigation towards the creation of low carbon transportation systems during the recent years, many studies have been used [Nanaki and Koroneos, 2017; Nanaki and Koroneos, 2016; Nanaki et al., 2014]. Some studies focused on the investigation of the energy efficiency, the environmental footprint as well as the environmental impacts of the urban transportation system, with special attention on road infrastructure [Huang et al., 2008; Taptich and Horvvath, 2014], vehicles [Nanaki and Koroneos, 2013] and transportation fuels [Yan and Crookes, 2009; Dalianis et al., 2016]. Nonetheless studies focused on the investigation of energy and exergy flows of the transportation sectors and their respective losses are limited. The above mentioned necessitate the investigation of energy and exergy flows and their respective losses, which will enable to address the goals of a sustainable transportation system by obtaining the optimum energy efficiency and by mitigating its consequent environmental impacts in the near future.

During the past decades a number of studies investigated the significance of optimizing a system by increasing its energy efficiency [i.e. Belnov et al., 2007; Saidur et al., 2010]. The results of these studies made clear that the term of energy efficiency cannot describe the performance of a system. As already mentioned in **Chapter 2**, exergy is a measure of the quality of energy that can be considered to evaluate, analyze and optimize the system. Exergy analysis is utilized to define the maximum performance of a system and to specify its irriversibilities [Yilansi et al., 2011]. The exergy function itself is an 'extensive' property to measure the effectiveness or real value of an energy form [Rosen and Dincer, 1997].

In this direction, exergy analysis has been employed, in order to assess the energy utilization of a country's energy system [i.e. Canada [Lemieux and Rosen, 1989; Terkovics and Rosen, 1988]; Japan, Finland and Sweden [Wall, 1990; Wall, 1991], UK [Hammond and Stapleton, 2001]. Studies regarding the analysis of **urban transportation systems** using exergy analysis have been performed for the countries of Canada [Rosen, 1992], Turkey [Ediger and Camdali, 2007; Utlu and Hepbasli, 2006; Seckin et al., 2013], Saudi Arabia [Dincer et al., 2004], Norway [Ertesvag, 2001], Jordan [Jaber et al., 2008], Malaysia [Saidur et al., 2007], UK [Gasparatos et al., 2009], Greece [Koroneos and Nanaki, 2008] and China [Chen et al., 2006]. Among these, none of them have investigated the interrelations of energy, exergy of the transportation sector and sustainability.

The high-energy costs of conventional energy resources and their environmental impact have raised the interest of increasing the energy efficiencies of urban transportation systems. Increasing efficiency usually reduces environmental impact and also has sustainability implications as it lengthens the lives of existing resource reserves. Although increasing efficiency generally entails greater use of materials, labor and more complex devices, the additional cost may be justified by the resulting benefits.

This chapter investigates the application of energy and exergy analyses to complex systems, such as countries, regions, and economic sectors as these are reflected in the transportation sector. Demonstrating the utilization of energy resources in society from the exergy point view is recommended and will result eventually in gaining additional knowledge on how to achieve a sustainable transportation system by applying measures to enhance efficiency. This chapter will also highlight the concept of energy and exergy and sustainability within the context of energy transition to a low carbon transportation system. A case study is provided for the energy and exergy utilization of Greece's transportation sector.

5.2 Urban Environment and Transportation Systems

Energy plays a significant role in the interrelations of economy and the natural as well as urban environment. Taking into consideration the fact that the energy content of an energy source is the available energy per unit of weight or volume, it is challenging to effectively extract and use this energy. Thus, the more energy consumed the greater the amount of work realized indicating by this way that economic development is correlated with greater levels of energy consumption.

Urban environments such as cities consume energy either directly or indirectly through the embodied energy in imported goods and services.

Transport energy consumption in urban areas varies from resource consumption in building materials (e.g. concrete for airport runways, asphalt for highways, and steel for railways) to fossil fuels to run different transportation modes [BTS, 2012]. Furthermore, population growth and economic expansion create significant problems to transportation systems. Urban transportation plays a great role to a city's sustainability, particularly in the case of emerging cities whose economies are developing rapidly alongside massive urban sprawl [Legates, 2014; Salama and Wiedmann, 2013]. For this reason, transportation systems should be able to support the economy, social activities as well as to contribute to environmental protection and resource effective utilization. In this context, the investigation and analysis of the way transportation systems operate and interrelate within the urban system, is of great significance for the creation of a zero carbon transportation sector.

The transportation sector accounts for 26% of worldwide energy consumption and 25% of direct and indirect carbon emissions [IEA, 2008]. Road transportation is responsible for the consumption of 85% of the total energy used in developed countries. Despite a falling market share, rail transport, based on 1 kg of oil equivalent, remains four times more efficient for passengers than and twice as efficient for freight movement as road transport. Rail transport accounts for 6% of global transport energy demand. Maritime transportation accounts for 90% of cross-border world trade as measured by volume. Air transportation plays an integral part in the globalization of transportation networks. The aviation industry accounts for 8% of the energy consumed by transportation. Further distinctions in the energy consumption of transport can be made between passenger and freight movements. Passenger transportation accounts for 60 to 70% of energy consumption from transportation activities. Freight transportation is dominated by rail and maritime shipping.

In regards to fuel consumption, worldwide, petroleum and other liquid fuels are the dominant source of transportation energy consumption. While the use of petroleum for other economic sectors, such as industrial and electricity generation has remained relatively stable, the growth in oil demand is mainly attributed to the growth in transportation demand. Transportation is almost completely reliant (95%) on petroleum products with the exception of railways using electrical power. It is to be noted that from 1971 to 2006 energy consumption in the transportation sector rose between 2.0 and 2.5% annually. The road transportation sector consumes the most energy followed by aviation. While in industrialized countries energy consumption has stabilized at a high or slightly declining level, the growth rate of transport energy consumption in non-OECD countries between 2000 and 2006 was 4.3% [IEA, 2009]. In addition, huge regional differences in transport energy use are noticed. The USA, Canada, Australia, European Union and Saudi Arabia are among the countries with the highest energy use per capita. In comparison India and neighboring countries as well as some parts of Africa use 20 times less transport energy per capita [IEA, 2009].

As far as the European Union is concerned, it is to be noted that in 2013, the transport sector accounted for 32% of final energy consumption in the EU-28, with road as the most important transportation mode (82.7%), followed by air (14.1%), railway (1.9%), and inland waterways (1.3%) [EC, 2015]. It is also to be noted, that the passenger – kilometers demand for passenger transport grew by more than 8% between 2000 and 2013 in the EU, with flying experiencing the most rapid growth. Furthermore, EU citizens traveled approximately 12 850 km per person in 2013 – more than 70% by car – representing a 5% increase since 2000. This growth indicates that road transport accounts for almost three quarters of the energy used in transport in the EU. Sales of new passenger cars in the EU increased by 9% in 2015 compared to the previous year, with 13.7 million new cars registered. Recent data indicate that diesel consumption in road transport, increased from 52% of total road fuel consumption in 2000 to 70% in 2014. Similarly, just over half of the vehicles sold in Europe are diesel, corresponding to 52% of sales in 2015. Finally, it is also noted that different types of electric vehicles (battery and/or combination of petrol/diesel hybrids) are also available on the European market.

Transport demand is closely linked to economic activity: in periods of growth, economic output goes up, more goods are transported and more people travel [Koroneos and Nanaki, 2008]. For instance, the economic recession of 2008 resulted in lower transport demand in the EU and, consequently, in reduced greenhouse gas emissions (GHG) from the sector in the years following. Despite this slow down period, the EU's overall transport emissions in 2014 were 20% higher than their 1990 levels [EEA, 2017].

5.3 Climate Change and Sustainable Transportation Systems

The effect of greenhouse gas emissions (GHG) on global temperature change and climate change is a challenging field for the sustainability of urban areas, such as cities. Transport is one of the sectors targeted where effective public interventions are being called for to reduce CO_2 emissions and where adaptation measures are needed to reduce the vulnerability to climatic changes. To be more specific, GHG emissions coming from the transport sector, have more than doubled since 1970, and increased at a faster rate than any other energy end-use sector to reach 7.0 Gt CO_2 eq in 2010 [IEA, 2012a]. It is pointed out, that the current CO_2 concentration in the atmosphere greatly exceeds the natural range (180–300 ppm) and this is expected to double (~660–790 ppm CO_2) by the end of this century [OECD, 2009].

In 2016, transportation accounted for the 28% of total U.S. GHG emissions [EPA, 2018]. Cars, trucks, commercial aircraft, and railroads, among other sources, all contribute to transportation end-use sector emissions. Within the sector, light-duty vehicles (including passenger cars and light-duty trucks) were by far the largest category, with 60% of GHG emissions, while medium- and heavy-duty trucks made up the second largest category, with 23% of emissions. Between 1990 and 2016, GHG emissions in the transportation sector increased more in absolute terms than any other sector (i.e. electricity generation, industry, agriculture, residential, or commercial) due in large part to increased demand for travel. In EU-28, transport contributed 24% to GHG emissions from all sectors in 2013 with transport GHG emissions 19% above levels in 1990 [EEA, 2015]. Passenger cars contributed to 44% of transport sector emissions, and heavy-duty vehicles and buses a further 18%. Emissions from different transport modes varied substantially over time. International aviation emissions almost doubled and road transport increased by 17% in this period, whereas emissions from rail transport and inland navigation declined by more than 50% and almost 37% respectively.

Based on the above mentioned, it is obvious that mitigating climate change depends strongly on the sustainability of energy systems. IPPC's report [IPPC, 2007] indicated the necessity of investigating all possible strategies for the reduction of GHG emissions. Stronger mitigation actions are required, so as to avoid greater impact from extreme climate change events [Stern, 2007; IPPC, 2007]. Nonetheless, the reduction of emissions in the transportation sector faces significant challenges due to its dependence on fossil fuels, to the high cost of low carbon technologies as well as to incomplete international agreements. For instance, the first global deal under the Kyoto Protocol, which came into effect in 2005, had a limited impact, as it only required developed countries to take action. What is more, the US never ratified it, Canada pulled out before the end of the first commitment period, which ended in 2012, and Russia, Japan and New Zealand did not take part in the second commitment period. As a result, the second commitment period of the Kyoto Protocol, from 2013 to 2020, only applies to around 14% of the world's GHG emissions-[www.consilium. europa.eu]. The second global deal for the climate change mitigation and adaptation is under the Paris Agreement, which was adopted in 2015 at the 21st session of the Conference of Parties (COP21) to the UNFCCC, in Paris. The deal committed developed and developing countries alike to keeping global warming below 2°C, and aspiring to a target of 1.5°C. The Agreement also has a long-term goal of net zero emissions entailing a complete decarbonization of the transport sector.

With regards to the EU, it is highlighted that the EU has set several targets to reduce GHG emissions from transport. In its White Paper [EC, 2011], the European Commission set a target of a 60% reduction from 1990 levels by 2050. Transport also needs to contribute to the EU's overall targets for GHG emissions reductions by 2020 and 2030. Part of the 2030 target will be achieved through the EU Emissions Trading Scheme (EU ETS). Although it includes emissions from aviation, other transport emissions are excluded. This means that with the exception of intra-EU aviation, the remaining transport modes will need to contribute to the 30% reduction effort for the sectors excluded [EEA, 2017-online] from the EU ETS.

In this direction, a set of legislative strategies aiming to meet the economic, social and environmental needs of a society are essential components of a transport policy. The key role of innovative technologies in ensuring sustainable, efficient and competitive mobility in urban areas should be reflected in an integrated transportation plan. The need for GHG mitigation and CO_2 abatement measures as well as the need for improved fuel efficiency in the transport sector is an important part of a low carbon transportation system. This, can be achieved through: Innovative vehicle technologies (i.e. advanced engine management systems and efficient vehicle powertrains); The use of sustainable biofuels (first, second as well as third generation fuels); Improved transport infrastructure together with Intelligent Transport Systems (ITS) to avoid traffic congestion and to foster the use of intermodal transport (road, rail and waterways); Consumer information ; Legal instruments (such as tax incentives for low carbon products and processes, taxation of CO_2 intensive products and processes, etc.).

5.4 Sustainable Transportation Systems – An Exergetic Approach

The need for the use of sustainable energy systems – as already mentioned in the previous section has been recognized globally. Urban areas, such as cities, are highly complex entities with increased resource utilization rates. The need to solve the problems associated with these (climate change, air pollution, etc) is crucial in maintaining a high quality of life in a sustainable way. Therefore, the creation of a sustainable transportation system should be the main goal that characterizes strategies and solutions for the mobility within urban areas.

Sustainable development requires the resources to be used efficiently. Exergy methods are essential in improving efficiency, which allows society to maximize the benefits it derives from its resources while minimizing the negative impacts (such as environmental damage). Greater efficiency in utilization allows such resources to contribute to development over a longer period. Increased efficiency reduces environmental impacts and resource requirements (energy, material, etc.) to maintain systems to harvest energy. In this direction, it can be suggested that exergy can play an important role in evaluating the use of green energies and technologies [Rosen et al., 2008]. Exergy is a measure of the departure of the state of a system from that of the environment. Therefore, exergy depicts in the most appropriate way the relation between the second law of thermodynamics and environmental impact [Dincer, 2002].

Thus, an important element for a sustainable transportation transition within urban areas is the exergy analysis. The exergy of an energy form or a substance is a measure of its quality or potential to cause change. This indicates that exergy could be an effective tool for the measurement of the environmental impact potential of a substance or an energy form. It is emphasized, that the concept of exergy can provide useful information in regards to the efficiency, the cost effectiveness as well as the sustainability of an energy system. Exergy losses should be minimized in order to achieve the targets of sustainable development. Measures aiming to reduce the exergy losses will lead to increased exergy efficiency, which in turn is associated with reduced environmental impacts and increased energy security in an environmentally acceptable way by reducing the emissions that might otherwise occur. In order to control air pollution resulted from the transportation sector, efficiency improvement actions often need to be supported by new vehicle technologies or fuel substitution.

An exergy efficient transport system implies lower energy use which is close to sustainability, especially if it leads to a reduction in fossil fuel consumption. The benefits of reduction in the usage of fossil fuel include inter alias, the reduction in the depletion of a finite resource, the reduction in the production of greenhouse gases, and the reduction in pollution released from the combustion of internal combustion engines. Furthermore, exergy analysis when combined with economics constitutes a useful tool for the systematic analysis and optimization of energy systems. In this direction, exergy is a powerful and effective tool for helping to achieve the targets of sustainable development in the transportation sector.

Exergy is correlated to the environmental impact of a transportation energy system, due to the fact that the pollution potential is proportional to the extent of energy conversion and utilization processes. The irreversibility of the real processes is synonymous to exergy destruction and waste flow to the environment. Therefore, the role of exergy analysis to the assessment of the sustainability of transportation systems is of great importance, mainly because exergy-based efficiency of systems and processes represents a true measure of their imperfections. For instance, the assessment – in terms of exergy analysis – of alternative energy technologies in the transportation sector such as hydrogen and fuel cell is important for achieving sustainability in both developing and industrialized countries. These technologies are a key component of sustainability in the transportation sector as they are energy efficient; compatible with renewable energy sources; they safeguard energy security, economic growth and sustainability. In addition, the performance of exergy analysis in such systems is essential in improving efficiency, allowing by this way society to maximize the benefits it derives from its resources while minimizing the environmental damage [Nanaki and Koroneos, 2017].

Figure 5.1 illustrates the relation between exergy, sustainability and environmental impact [Rosen et al., 2008]. It can be discerned that by increasing exergy efficiency the sustainability increases and the environmental impact decreases. An energy system with 100% exergy efficiency is completely reversible and has no environmental impact and therefore sustainability approaches to infinity. On the other hand, a system with zero exergy efficiency has significant environmental impact.

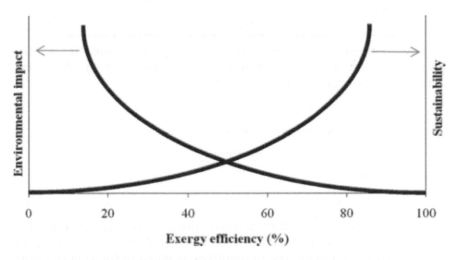

Fig. 5.1. Qualitative illustration of the relation between exergy efficiency, environmental impact and sustainability [Rosen et al., 2008]

5.5 Exergetics of Transportation Systems

In order to evaluate the exergy flows associated with a vehicle, the standard atmosphere is considered as the reference environment [Gaggioli and Petit, 1977]. The physical exergy, for fuels used in transportation devices, is negligible when compared to the chemical exergy, which is estimated as [Kotas, 1985]:

$$\varepsilon = \gamma * LHV \tag{1}$$

where ε stands for the specific exergy, γ denotes the exergy factor based on Lower Heating Value (LHV), which is the energy input for the considered process. Table 5.1 shows related data for selected fuel forms.

Table 5.1. Exergy analysis related data for selected fuel forms [Kotas, 1985]

Fuel forms	LHVs (kJ/kg)	Exergy factors
Crude oil	41,816	1.08
Gasoline	43,070	1.06
Diesel oil	42,652	1.07
Kerosene	43,070	1.07
Fuel oil	41,816	1.06
LPG	50,179	1.06
Other petroleum products	42,000	1.06
Coal	26,344	1.08

The expressions of **energy (n)** and **exergy (ψ) efficiencies** for the principal types of processes considered are based on the following definitions. Energy efficiency is defined as:

$$n = \text{work/energy input} \tag{2}$$

whereas exergy efficiency is defined as:

$$\psi = \text{work/exergy input} \tag{3}$$

it is obvious that:

$$\psi = n/\gamma \tag{4}$$

the exergy efficiency is equal to the conventional energy efficiency divided by the exergy factor. It should be mentioned that the exergy value, for the electricity used in electric locomotives, is equal to the energy value; based on this reasoning, the exergy efficiency is the same as the energy efficiency. The weighted mean overall exergy efficiency is calculated as:

$$\psi_{overall} = \sum_{i,k}\left(\frac{n_i}{\gamma_k}\right) \times Fr_{ik} \tag{5}$$

where ψ overall expresses the weighted mean overall exergy efficiency, ni stands for the energy efficiency of the ith transportation mode, γk the exergy factor of the kth energy form and Fr_{ik} denotes the exergy fraction of the kth energy form used by the ith transportation mode.

Table 5.2 highlights the distinction between energy and exergy efficiencies for selected processes. The energy efficiencies stand for the energy of the useful streams leaving the process divided by the energy of all entering streams. The exergy efficiencies stand for the ratio of the exergy contained in the products of a process to the exergy in all input streams. As already mentioned in Chapter 2, Table 5.2 indicates that the exergy efficiencies are lower than the energy efficiencies; this is attributed to the destruction of the input exergy due to irreversibilities. Out of these reasons the exergy efficiency frequently gives a finer understanding of performance than the energy efficiency.

Table 5.2. Energy and exergy efficiencies for selected processes
[Rosen and Dincer, 1997]

Process	*Energy efficiency (%)*	*Exergy efficiency (%)*
Petroleum refining	~90	10
Residential heater (fuel)	60	9
Domestic water heater (fuel)	40	2–3
Coal gasification (high heat)	55	46
Steam-heated reboiler	~100	40
Blast furnace	76	46
High-pressure steam boiler	90	50

5.6 Case Study: Exergetics of the Greek Transportation System

The application of the methodology discussed in Section 5.4 is presented for the energy and exergy use in the transport sector of Greece. The transportation sector of Greece consists of four main modes, namely highways railways, marine and civil aviation. The energy and exergy efficiencies are calculated by multiplying the energy used in each end use by the corresponding efficiency for that end use. The overall efficiency of the transportation sector was obtained by adding these values. Furthermore, a weighted mean is obtained for the transportation mode energy and exergy efficiencies, where the weighting factors are the total energy and exergy inputs, which supply to each transportation mode.

The energy efficiency data for the four modes under consideration are obtained from Reistad (1975). These data are based on US devices and are assumed representative of the devices used in Greece. Since vehicles are not operated at full load, part load efficiencies of their devices are estimated as 22, 28, 15 and 28% for road, rail, marine and air, respectively. Based on the methodology presented in Section 5.4 and the data of Table 5.3 the overall exergy efficiency for the transportation sector in the year 2000 is calculated as per Eq. (6):

$$\psi_{overall} = (22\%/1.06) \times 0.4677 + (22\%/1.07) \times 0.2672 + (22\%/1.06) \times 0.0022$$
$$+ (28\%/1.00) \times 0.0014 + (28\%/1.07) \times 0.0056$$
$$+ (15\%/1.08) \times 0.0312 + (15\%/1.07) \times 0.0371$$
$$+ (28\%/1.07) \times 0.1876 = 21.23\% \tag{6}$$

Table 5.3. Calculation of the energy and exergy efficiency of
the Greek transport sector in 2000

Year	Transportation mode	Fuel	Energy (PJ)	Exergy (PJ)
		Gasoline	144,67	153,35
	Highways	Diesel	81,88	87,61
		LPG	0,712	0,75
	Railways	Electricity	0,335	0,335
2000		Diesel	1,72	1,84
	Marine	Fuel oil	9,5	10,26
		Diesel	11,39	12,19
	Civil Aviation	Jet	57,51	61,53
				327,865

With regards to the biodiesel, which was introduced to the Greek transportation system in 2005, it is noted that the relevant exergy coefficient is calculated based on the Extended Exergy Analysis (EEA) performed [Peiró et al., 2010] for the production of 1 tone of biodiesel from rapeseed oil in Catalonia in South Europe.

Biodiesel is a renewable fuel produced from vegetable oils such as rapeseed and sunflower seed oils, and also from Used Cooking Oil (UCO). UCO is a mixture with high content of tri-, di- and monoacylglycerides, and Free Fatty Acids (FFA) [Knothe et al., 2005]. The reaction between glycerides and short chain alcohols, such as methanol and ethanol, yields glycerol and fatty acids methyl esters, better known as biodiesel. The production of biodiesel is based on a two-step process. The first step is the preesterification of FFA and glycerides with an alcohol using an acid catalyst, such as sulfuric acid or phosphoric acid. The ester phase containing methyl esters and glycerides settles in the reactor and the unreacted FFA is fed to a separate tank. The second step is the transesterification of the glycerides with alcohol in the presence of an alkali catalyst (potassium hydroxide or sodium hydroxide) or an alkoxide (metoxide or etoxide). The reaction mixture settles and separates into an ester phase and a glycerol phase. The ester phase contains glycerides, methyl esters and methanol which are purified by distillation to obtain the final biodiesel. The remaining unreacted glycerides are reintroduced in the transesterification reactor together with traces of esters which allow for a better mixture of the alcohol and oil phases. The glycerol phase which contains glycerol, water and methanol, is fed into a buffer tank which also contains the glycerol phase from the transesterification reactor. Once the two glycerol phases are well mixed, they are fed into an acidulation tank where FFA from the preesterification are added until having an alkaline pH to avoid the formation of soaps and emulsions. The remaining unreacted FFA is sent again into a FFA buffer tank to be reused in the preesterification reactor. The glycerol phase is neutralized and distilled to recuperate glycerol and methanol to be reused within the system. By this two step process, the transesterification takes place at moderate conditions and the biodiesel conversion reaches its highest rate, above 95% [Mittelbach and Remschmidt, 2004].

Based on the above mentioned procedure as well as on Eq. 1 and taking into consideration the fact that the lower calorific value of rapeseed biodiesel is 35 MJ/kg [Coniglo et al., 2013], the exergy coefficient for the biodiesel is calculated as 2,20. As the biodiesel used in the Greek transport sector is a B5 blend (95% diesel and 5% biodiesel) and based on the fact that the exergy coefficient of diesel is 1,07 the exergy efficiency of B5 biodiesel blend is calculated as 1.13.

5.7 Results

Total energy and exergy efficiency of the Greek transportation system, during the period of 1980–2016 is presented in Fig. 5.2. The **energy efficiency** of the Greek transportation sector declined from 22.96 % in 1980 to 22.41% in 2004, 22.33% in 2011 and 22.18% in 2016. The exergy efficiency of the Greek transport sector presents a slowly decreasing trend from 21.46% in 1980 to 20.92% in 2003. It is noted, that the exergy efficiency remained almost stable till 2016 (20.94%). Since the exergy efficiencies of devices remained unchanged, the decrease of the overall exergy efficiencies could be attributed to the transition of the energy use structure in general. Furthermore, it is clearly seen that energy efficiencies are much higher than the corresponding exergy efficiencies, due to the fact that exergy considers the losses attributed to irreversibilities. It becomes obvious that the real overall efficiency of the Greek transportation sector is better depicted by exergy and not energy.

The time variation of total exergy consumption in the Greek transport sector for the period from 1980 to 2016 is shown in Fig. 5.3. The total exergy consumed in 2016 was 300,86 PJ, in 2003 was 352 PJ whereas in 1980 was 180 PJ. The ascending trends of the exergy consumption in Fig. 5.3 illuminate the increasing demand for mobility in Greece over the period 1980–2008. Nonetheless, the descending trends of exergy consumption especially in 2010 (363.90 PJ from 411.08 PJ in 2009) emphasize the impacts of economic crisis to the Greek transportation sector.

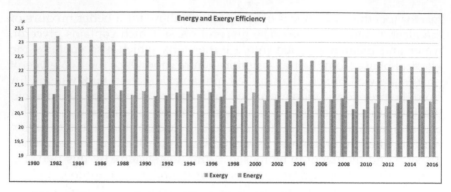

Fig. 5.2. Energy and exergy efficiency for the Greek transportation sector
during the period of 1980–2016

The annual budget deficit for the year 2009 for Greece (amounting to some 15% of the GDP for that year) resulted in an unprecedented economic crisis for the country and caused its inability to borrow money from the markets. As a consequence, a lending mechanism that was put up for the first time in a Euro zone country was introduced. This consisted of three lending parties i.e. the European Commission, the Central European Bank and the International Monetary Fund (IMF). The above mentioned was reflected to the total turnover per transport sector, which recorded a considerable drop between 2009 and 2013. These drops were most dramatic in 2009 reaching –31.5% for the inland transport sector [Hellenic Statistical Authority, 2015]. To be more specific, gasoline demand decreased by 1.9% during the first seven months of 2016, while diesel sales for the same period increased by 3.5%. However, demand for both products increased in July alone year over year, with gasoline and diesel demand increasing by 8.4 and 10.4%, respectively. It should be noted, that in July 2015 capital controls were imposed, which affected consumers' willingness to spend and, consequently, demand for fuel products dropped [Nanaki, 2018]. The demand for petroleum products in the Greek market, after peaking in 2007, has fallen dramatically, reversing a longstanding upward trend, which was closely connected with the growth of the Greek economy. Diesel fuel consumption by passenger cars was virtually increased after 2010, given that circulation of diesel passenger cars was prohibited in the two major metropolitan areas of Athens and Thessaloniki.

It is obvious that the transport sector not only plays a crucial role to socio-economic development but is also influenced both directly and indirectly by other socioeconomic factors such as fiscal policy, industry structure, etc. In particular, the growth of GDP, in Greece, during the period of 1980–2006 has led to an increase of the energy use of transport, reflecting increases in traffic and leading to increase in emissions of greenhouse gases. In addition, the fluctuations of fuel prices have affected not only the cost of driving, which is directly linked to fuel consumption, but all the sectors of the economy. It is obvious that as transport grows, it demands more energy. Therefore, transport demand and energy use are closely linked [Koroneos and Nanaki, 2008].

The exergy consumption of each mode for the period from 1980 to 2016 is presented in Fig. 5.4. In 1980, highways transport was the biggest exergy consumer with a share of 58%, and civil aviation, waterways and railways ranked second, third and fourth making up 29, 12 and 1% of the total respectively. The same exergy consumption trends are noticed in 2003, where highways transport was the biggest exergy consumer with a share of 75%, and civil aviation, waterways and railways ranked second, third and fourth making up 15, 9 and 1% of the total respectively. The same trends are also noticed in 2016, with highways transport being the biggest exergy consumer with a share of 75% and civil aviation, waterways and railways ranked second, third and fourth making up 16, 8 and 1% of the total respectively.

It is evident that the exergy consumed by highways increased rapidly over the period of 1980–2008. The time variation of the exergy consumption of each mode reflects changes of the energy utilization structure of the Greek transportation sector, which is influenced by socio-economic factors, policies, residents' attitude as well as the very characteristics of vehicles themselves. The demand for flexibility in conjunction with the

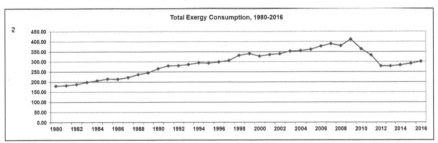

Fig. 5.3. Total exergy consumption of the Greek transportation system

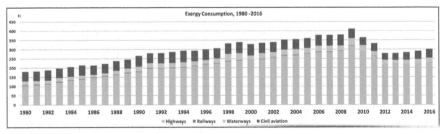

Fig. 5.4. Exergy consumption of the subsystems of the Greek transportation sector

demand for convenience as well as the economic development plays a crucial role in highway transportation. Details of the exergy consumption of the highways subsector for the period from 1980 to 2016 are shown clearly in Fig 5.5.

The road subsector appears to be the most energy and exergy efficient one. This could be attributed to the fuel type used and the performance of the carrier. It should be pointed out, that since natural gas has the lowest exergy factor (1.04), it gives the highest energy and exergy efficiency compared to other fuels listed in **Table 5.1.** Thus, the use of natural gas in the transport sector is beneficial to the system's efficiency. The latter is of great significance, as the environmental impacts are going to be further minimized.

In this direction, the promotion and penetration of alternative and renewable transport energy sources, such as natural gas, biofuels and electricity should be further investigated in the transport policy. A shift towards cleaner fuels such as unleaded petrol and low-sulfur diesel, LPG, methane and non-fossil energy sources is expected to lessen the environmental impacts of the transport sector. Greece has shown a great interest in this area as the uptake of cleaner fuels over the period 1990–2016 has increased, especially in 2006 after the introduction of biofuels and in 2012 after the introduction of electromobility. What is more, changes in the global energy status in conjunction with global demand and European directives will have some significant effects in the transport sector; therefore, the efficiency of renewable fuels in the transportation sector should be further investigated. Finally the multi-year (1980–2016) average overall energy and exergy efficiencies of each transport mode in Greece reveal that the road subsector appears to be the most energy and exergy efficient one. This could be attributed to the fuel type used and the performance of the carrier.

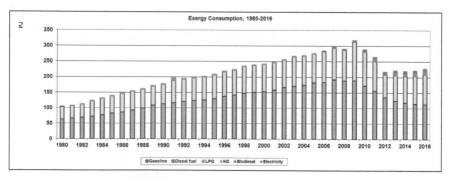

Fig. 5.5. Exergy consumption of Greek transportation system – analysis per fuel consumption in road subsector

5.8 Concluding Remarks

This chapter examined the role of the exergy concept in climate change adaptation and mitigation within urban areas such as cities as well as its contribution to a creation of a sustainable transportation system.

It was suggested that a sustainable transportation system refers to a balanced pursuit of multiple objectives, taking into account the impacts on local society, economy, and the environment. The assessment of the sustainability of transportation systems is beneficial to decision–makers as it can be used as a guide to future investment of a renewable energy source based transportation system. The method of exergy analysis can be applied, in order to enhance sustainability. The use of exergy analysis is one important element in obtaining sustainable development. Environmental effects associated with emissions and resource depletion can be expressed in terms of exergy.

The main stakeholders in a transportation system are grouped in society, economy and environment. It is of great significance that the interests of these groups are all integrated into the transportation system as they are all essential parameters in society. Exergy analysis enables constructive communication between different professions. The lack of a holistic and flexible focus in the planning strategies will inevitably lead to an unsustainable transport system; therefore, exergy analysis is beneficial as a diagnostic tool in the stage of designing a transportation system.

The case study applied the method of exergy analysis in the energy and exergy utilization of the Greek transportation sector, based on actual data, by considering the energy and exergy flows for the years of 1980–2016. The variations of energy and exergy efficiencies for the transportation sector were studied for its four subsectors, namely highways, railways, marine and civil aviation. From the analysis it is shown that taking into consideration the expanding demand for mobility, the total exergy consumed by the transportation sector during the period of 1980–2008 increased significantly; whereas during the period of 2008–2016 the total exergy decreased. This was mainly attributed to the economic crisis. Furthermore, the road subsector appears to be the most efficient one compared to the railways, marine and civil aviation.

Exergy analysis and its corresponding results offer constructive suggestions for the optimization and improvement of the transport system since it provides a linkage between the physical and engineering world and the surrounding environment. Because of this reason such an analysis can be used in a sustainable transportation planning.

REFERENCES

Alises, A. and Vassallo, J.M. (2015). Comparison of road freight transport trends in Europe. Coupling and decoupling factors from an Input-Output structural decomposition analysis. Transportation Research Board 82: 141–157.

Belnov, V.K., Voskresenskii, N.M. and Kheifets, L.I. (2007). Increasing the thermodynamic efficiency of an air-separating plant by recuperating the mechanical energy of the gas streams. Theoretical Foundations of Chemical Engineering 41(5): 519–525.

BTS, Economic impact on transportation (2012). (O. o. Technology, Editor & U.S. Department of Transportation) Retrieved February 14, 2015, from Bureau of Transportation Statistics: http://www.rita.dot.gov/bts/programs/freight_transportation/html/transportation.html

Chen, B., Chen, G.Q. and Yang, Z. (2006). Exergy-based resource accounting for China. Ecological Modelling 196(3–4): 313–328.

Coniglio, L., Bennadji, H., Glaude, P.A., Herbinet, O. and Billaud, F. (2013). Combustion chemical kinetics of biodiesel and related compounds (methyl and ethyl esters): Experiments and modeling – Advances and future refinements. Progress in Energy and Combustion Science 39(4): 340–382.

Dalianis, G., Nanaki, E., Xydis, G. and Zervas, E. (2016). New aspects to greenhouse gas mitigation policies for low carbon cities. Journal of Energies, MDPI Publishers 9(3): 128.

Dincer, I., Hussain, M.M. and Al-Zaharnah, I. (2004). Energy and exergy utilization in transportation sector of Saudi Arabia. Applied Thermal Engineering 24(4): 525–538.

Dincer, I. (2002). The role of exergy in energy policy making. Energy Policy 30(2): 137–149.

Ediger, V.S. and Camdalı, U. (2007). Energy and exergy efficiencies in Turkish transportation sector, 1988–2004. Energy Policy 35(2): 1238–1244.

Environmental Protection Agency – EPA (2018). US Transportation Sector Greenhouse Gas Emissions, 1990–2016, July 2018.

Ertesvag, I.S. (2001). Society exergy analysis: A comparison of different societies. Energy 26(3): 253–270.

European Commission – EC (2015). EU Transport in Figures. Statistical Pocketbook 2015. http://ec.europa.eu/transport/facts-fundings/statistics/pocketbook-2015_en.htm

European Commission – EC (2011). Roadmap to a Single European Transport Area – Towards a Competitive and Resource Efficient Transport System. Luxemburg, 2011.

European Environment Agency – EEA (2016). Transport in Europe: Key Facts and Trends: https://www.eea.europa.eu/signals/signals-2016/articles/transport-in-europe-key-facts-trends/#footnote3 (accessed on September 2018).

European Environment Agency – EEA (2017): https://www.eea.europa.eu/signals/signals-2016/articles/transport-in-europe-key-facts-trends/#footnote1

European Environment Agency – EEA (2015). Evaluating 15 years of transport and environmental policy integration. TERM 2015. Copenhagen. http://www.eea.europa.eu/ publications/term-report-2015.

Fuglestvedt, J., Berntsen, T., Myhre, G., Rypdal, K. and Skeie, R.B. (2007). Climate forcing from the transport sectors. Proceedings of National Academy of Sciences of the United States of America 105: 454–458.

Gaggioli, R.A. and Petit, P.J. (1977). Use the second law first. CHEMTECH (United States) 7: 8.

Gasparatos, A., El-Haram, M. and Horner, M. (2009). A longitudinal analysis of the UK transport sector, 1970–2010. Energy Policy 37(2): 623–632.

Geng, Y., Ma, Z., Xue, B., Ren, W., Liu, Z. and Fujita, T. (2013). Co-benefit evaluation for urban public transportation sector – A case of Shenyang, China. Journal of Cleaner Production 58: 82–91.

Hammond, G.P. and Stapleton, A.J. (2001). Exergy analysis of the United Kingdom Energy System. IMech – Journal of Power and Energy, 215: 141–162.

Hellenic Statistical Authority (2015). Greece in Figures. July-September 2015 Piraeus.

Huang, Y., Bird, R. and Bell, M. (2008). A comparative study of the emissions by road maintenance works and the disrupted traffic using life cycle assessment and micro-simulation. Transportation Research Part D: Transport and Environment 14: 197–204.

International Energy Agency – IEA (2008). Tracking Industrial Energy Efficiency and CO_2 Emissions (Paris, France), 324.

International Energy Agency – IEA (2009). Transport, Energy and CO_2 – Moving toward Sustainability (Paris, France: OECD/IEA, 2009).

International Energy Agency – IEA (2012). World Energy Outlook 2012.

International Energy Agency – IEA (2012a) CO_2 Emissions from Fuel Combustion Beyond 2020 Online Database.

International Energy Agency – IEA (2014). Key World Energy Statistics 2014.

IPCC Climate Change (2007). Fourth assessment report. Working Groups I, II, and III. World Meteorological Organization and the United Nations Environment Programme. Cambridge University Press, Available at: http://www.consilium.europa.eu/en/policies/climate-change/international-agreements-climate-action/.

Jaber, J.O., Al-Ghandoor, A. and Sawalha, S.A. (2008). Energy analysis and exergy utilization in the transportation sector of Jordan. Energy Policy August 36(8): 2995–3000.

Ji, X. and Chen, G.Q. (2006). Exergy analysis of energy utilization in the transportation sector in China. Energy Policy 34(14): 1709-1719.

Knothe, G., Van Gerpen, J. and Krahl, J. (Eds.) (2005). The Biodiesel Handbook. AOCS Press. Champaign, Illinois.

Koroneos, C.J. and Nanaki, E.A. (2008). Energy and exergy utilization assessment of the Greek transport sector. Journal of Resources, Conservation and Recycling 52(5): 700–706.

Kotas, T.J. (1985). The Exergy Method of Thermal Plant Analysis. Elsevier, ISBN 978-0-408-01350-5

Legates, R.T. (2014). Visions, scale, tempo, and form in China's emerging city-regions. Cities 41: 171–178.

Lemieux, M.A. and Rosen, M.A. (1989). Energy and exergy analysis of energy utilization in Ontario. Research Report. Ryerson Polytechnic University, Toronto.

Mathiesen, B.V., Lund, H., Connolly, D., Wenzel, H., Ostergaard, P.A. and Möller, B. (2015). Smart energy systems for coherent 100% renewable energy and transport solutions. Applied Energy 145: 139–154.

Mittelbach, M. and Remschmidt, C. (2004). Biodiesel: The Comprehensive Handbook. 1st ed. Austria: Martin Mittelbach.

Nanaki, E.A. and Koroneos, C.J. (2013). Comparative economic and environmental analysis of conventional, hybrid and electric vehicles – The case study of Greece. International Journal of Cleaner Production, Elsevier 53: 261–266. ISSN: 0959-6526.

Nanaki, E.A. and Koroneos, C.J. (2016). Climate change mitigation and deployment of electric vehicles in urban areas. Journal of Renewable Energy 99: 1153–1160.

Nanaki, E.A. and Koroneos, C.J. (2017). Exergetic aspects of hydrogen energy systems – The case study of a fuel cell bus. Journal of Sustainability, MDPI Publishers 9(2): 276.

Nanaki, E.A., Koroneos, C.J., Xydis, G.A. and Rovas, D. (2014). Comparative environmental assessment of Athens urban buses – diesel, CNG and biofuel powered. Journal of Transport Policy, Elsevier Publishers 01/2014.

Nanaki, E. (2018). Measuring the impact of economic crisis to the Greek vehicle market. MDPI Sustainability 10: 510.

OECD Reducing transport GHG emissions: Opportunities and costs (Online). Available from International Transport Forum (2009). http://www.internationaltransportforum.org/Pub/pdf/09GHGsum.pdf

Peiró, L. Talens, Villalba Méndez, G., Sciubba, E. and Gabarrell i Durany, X. (2010). Extended exergy accounting applied to biodiesel production. Journal of Energy 35(7): 2861–2869.

Reistad, G. (1975). Available Energy conversion and utilization in the United States. Journal of Energy Power 97: 429–434.

Rosen, M.A., Dincer, I. and Kanoglu, M. (2008). Role of exergy in increasing efficiency and sustainability and reducing environmental impact. Energ Policy 36(1): 128–137.

Rosen, M.A. and Dincer, I. (1997). Sectoral energy and exergy modeling of Turkey. ASME Journal of Energy Resources Technology 119(3): 200–204.

Rosen, M.A. (1992). Evaluation of energy utilization efficiency in Canada using energy and exergy analyses. Energy 17(4): 339e50.

Saidur, R., Ahamed, J.U. and Masjuki, H.H. (2010). Energy, exergy and economic analysis of industrial boilers. Energy Policy 38(5): 2188–2197.

Saidur, R., Masjuki, H.H. and Jamaluddin, M.Y. (2007). An application of energy and exergy analysis in residential sector of Malaysia. Energy Policy 35(2): 1050-1063.

Salama, A.M. and Wiedmann, F. (2013). The production of urban qualities in the emerging city of Doha: Urban space diversity as a case for investigating the 'lived space'. Environmental and Molecular Mutagenesis, Wiley 21: 180.

Seckin, C., Sciubba, E. and Bayulken, A.R. (2013). Extended exergy analysis of Turkish transportation sector. Journal of Cleaner Production 70: 422–436.

Stern, N. (2007). The Economics of Climate Change. Stern Review on the Economics of Climate Change. Cambridge University Press, Cambridge.

Talens, L., Peiró, Villalba Méndez, G., Sciubba, E. and Gabarrell i Durany, X. (2010). Extended exergy accounting applied to biodiesel production. Journal of Energy, Elsevier 35: 2861–2869.

Taptich, M.N. and Horvath, A. (2014). Bias of averages in life-cycle footprinting of infrastructure: Truck and bus case studies. Environmental Science & Technology, ACS Publications 48: 13045–13052.

Terkovics, P.J. and Rosen, M.A. (1988). Energy and Exergy Analysis of Canadian Energy Utilization. Research Report. Ryerson Polytechnic University, Toronto.

U.S. Energy Information Administration – EIA (2016). International Energy Outlook 2016.

United Nations (2007). World Urbanization Prospects: The 2007 Revision. New York.

Utlu, Z. and Hepbasli, A. (2006). Assessment of the energy utilization efficiency in the Turkish transportation sector between 2000 and 2020 using energy and exergy analysis method. Energy Policy 34(13): 1611-1618.

Wall, G. (1990). Exergy conversion in the Japanese society energy. Energy 15(5): 435–444.

Wall, G. (1991). Exergy conversions in the Finish, Japanese and Swedish societies, opuscula exergy papers. University College of Eskilstuna/Vasteras, Sweden, pp. 1–11.

Yan, X. and Crookes, R.J. (2009). Reduction potentials of energy demand and GHG emissions in China's road transport sector. Energy Policy 37: 658–668.

Yilanci, A., Ozturk, H.K., Dincer, I., Ulu, E.Y., Cetin, E. and Ekren, O. (2011). Exergy analysis and environmental impact assessment of a photovoltaic-hydrogen production system. International Journal of Exergy 8(2): 227–246.

Exergy Analysis of Intelligent Energy Systems in the Built Environment

G. Xydis[1] and E. Nanaki[2]

[1] Department of Business Development and Technology, Aarhus University, Birk Centerpark 15, 7400 Herning, Denmark
[2] University of Western Macedonia, Department of Mechanical Engineering, Bakola & Sialvera, Kozani 50100, Greece

6.1 Introduction

The building sector in the EU is responsible for about 40% of total energy consumption and for about 36% of total CO_2 emissions. Over the last decade, the EU has adopted measures to promote the rational use of energy including new legislation and new proposed methodologies. One of the most important energy policy initiatives is the recast of the Directives on "Energy in Buildings Performance" [Directive 2010/31/EU] and on "Energy Efficiency" [Directive 2012/27/EU], which state that Member States should set national energy efficiency requirements in a harmonized calculation methodology with a view to achieving cost-optimal levels, and set a general framework for national-wide energy efficiency action plans.

Inefficiencies, irreversibilities, and exergy losses within the built environment have always been there and since the urban population is increasing those are increasing as well. In India for instance, the world's fastest growing major economy, the 10 fastest growing cities by GDP are found. It is estimated that Delhi will add 10 million more people by 2030, and by 2050 68% of India's population will live in cities (Fig. 6.1).

A whole new thinking with regards to establishing innovative exergetic indicators and methodologies, is clearly needed within cities [Koroneos et al., 2012]. The value of these proposed indicators should describe energy losses (e.g. heating losses), and define the overall impact

Fig. 6.1. World Economic Forum (https://www.weforum.org/) stresses that India faces an unprecedented rate of urbanization

Color version at the end of the book

towards a "smarter city" or a "smart city v2.0". The use of "exergy indicators" based on the exergy content of the processes should present the relationship between energy utilized and energy wasted illustrating in this way the environmental impact.

The goal of this analysis is to provide a deeper analysis and guidance on the proper application of the cost-optimal methodology to the built environment and assess the impact of the various critical parameters. For instance, what are the principles today of the bioclimatic design and the use of renewable energy sources in the buildings? What are the cost-optimal levels of a house e.g. for a 30 year period? What could be the specific energy efficiency measures that could improve e.g. the thermal insulation of the building envelope, the operation of the technical systems, the optimal combination of energy related measure, such as cold storage applications [Xydis, 2013a] and demand response programs for smart communities [Xydis et al., 2013]? What is clearly disappointing is that even in developed and high-tech countries, the ratio of electricity and energy losses is increasing. Heating is done via electricity or electricity is still at times state-regulated (not fully liberalized electricity markets). That means that the cost of the average yearly electricity marginal price cost could reach up to 40-55 EUR/MWh, while in more mature – and completely liberalized – electricity markets this cost could fall at 30 EUR/MWh or even lower, which is a huge difference within the EU. Even market coupling – which is obligatory for some hours e.g. throughout the day according to the EU "Target Model" – and interconnected markets cannot polish the turd.

Numerous studies have dealt with cost optimization [Arribas et al., 2010; Vairavamoorthy et al., 2008], maximizing levels of thermal insulation of the building – and minimizing losses – via various energy sources [Dombayci et al., 2006], CO_2 decrease studies [Hirano and Fujita, 2016] air quality studies [Perini et al., 2011; Xydis, 2012a], the urban heat island phenomenon and albedo [Susca, 2012], green roofs [Zinzi and Agnoli, 2012; Jim and Tsang, 2011; Susca et al., 2011], photovoltaic systems [Xydis, 2013b], renewable energy sources [Matic et al., 2016; Vardopoulos, 2018; Xydis, 2012b], urban transportation systems [Nanaki et al., 2017] and so many other urban related topics, however always independently. Modern societies are at this point where all urban related topics are to be integrated under one platform. This platform will evaluate the source of energy, the use of it, according to the demand and will determine irreversibilities at a district level. Exergy analysis is a strong tool that can be utilized to a wider extent. So far, most of its usage is on specific products, processes mainly due to the complexity of the systems and the information required to be gathered.

The oil depletion crisis has led to a different way of thinking regarding natural resources. Nevertheless, the abstract way of dealing with the natural resources and their usage did not help towards climate change and eventually climate crisis. Renewable Energy Sources (RES) contributed and contribute to the overall solution of the 2 degrees problem however the time for reverting the problem and its impacts on human and society is not enough.

Over the last 20 years, replacing the existing autonomous thermal power plants by Distributed Energy Resources (DER) has become a first priority matter for a number of municipalities, system operators, and local communities. DER using renewable energy technologies, along with appropriate energy storage units was considered as the weapon to deal with the major problems that the non-interconnected islands or isolated areas face, aiming to meet at the same time the national energy targets.

However, dealing with the minimization of various energy and water shortages, at times, and delivering studies conducted for assessing the feasibility of DER deployment cannot solve the problem of inefficient networks. It is necessary for all – or at least most of the – city's processes to be optimized (Fig. 6.2).

But this is a long lasting – and unclear task for for the continuously changing urban environment. If a city/region alters its energy system to a greener one and via this way minimizes the CO_2 production but on the other hand continues to import fresh vegetables and leafy products in large quantities from distant areas the local CO_2 result might not change

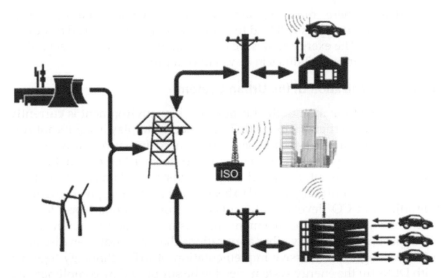

Fig. 6.2. A bi-directional communication system for the "smart city 2.0"

significantly. What is needed is a holistic optimization approach which will serve citizens, society, sustaining natural resources.

The main objectives should be cost minimization, CO_2 emissions minimization, DER reliability maximization, environmental effects at all levels for all stakeholders aiming at mutual beneficial solutions. There are a great number of decision support tools used for multi-objective analysis of resources, which propose a set of optimal solutions defining the appropriate technologies and capacities of these technologies.

The main objective for city planners and decision makers is to evaluate the urban system of each city. To achieve this goal, Geographic Information Systems (GIS) should be utilized and energy usage, emissions data, plants' operating data, water network losses, food transport routes, waste management activities, commercial and tourist activities should be registered for the metropolitan areas. This seems as a humongous task, however it is absolutely necessary for the overall efficiency of urban areas. And since the aim is to minimize the total losses within a nexus (e.g. a water-food-waste-energy nexus), exergy, as a system evaluator, could play a dominant role.

6.1.1 Exergy is a Concept for Resources Degradation

Exergy is a concept that shows the quality of energy and matter, in addition to what has been consumed in the course of energy transfer or conversion steps. The concept of "exergy" provides the information for further understanding of "how a system works" by pinpointing the subsystems where resources are minimized. Understanding exergy consumption principles will lead to a better understanding of all resources management. The exergy analysis can accurately indicate the exact points and causes of irreversibilities in a system or a process.

6.1.2 Modernization of the Urban System

A major transformation of the European urban environment is currently taking place. Internet of Things (IoT) is expected to play a significant role in the near future. IoT technology can be used in various ways within cities and investigate citizens' needs related to smart cities, what type of services do they want, need and why. For instance, data collection from apartments and buildings related to heating consumption, power, lighting, humidity and CO_2 levels can be collected from indoors. Noise level in cities, emissions, root management for waste collection only when bins are almost full and many other applications that were not even discussed 5-10 years ago. The massive implementation of IoT technology together with DERs (in the energy system) need to be supported by complimentary technologies for intelligent control. Advanced systems are required to maintain or increase system robustness, stability and security. Urban areas are of particular interest due to their dynamics and due to scarcity of resources and at the same time increased demand. Urban areas account for approximately 70% of the primary EU energy consumption and the electricity share is expected to increase significantly with the integration of electric vehicles.

There is currently a lack of understanding of the aggregated impact of the local systems and components. What is the overall performance of an integrated urban system? Researchers, urban planners and technology developers require appropriate methods and tools for analyzing, visualizing and optimizing the performance of all the multi-components urban systems predominantly based on acquiring all the wasted city data.

Such a holistic approach and effort has not yet been implemented, at least in an organized manner. Researchers and urban planners from all around the world work on specific challenges and not on integrating all urban challenges under one scheme, one platform.

6.2 Artificial Intelligence and The Urban Nexus Platform

What is needed worldwide in the built environment is to establish a unique large scale simulation platform with a selection of tools for urban community energy systems simulations and testing. The Urban Nexus Platform (UNEX) will serve as a central tool for developing, testing, and practically implement distributed micro energy resources and their control systems in future smart cities by quantifying the overall performance and revealing the impacts into their full energy system. The platform will help to identify and optimize the true potentials of new energy technologies and avoid unforeseen risks and side effects.

The platform will be the base for the exergetic-based analysis of all the nexus components, being valuable and guiding numerous urban energy projects for new or existing city areas establishing well-defined energy performance targets and measures. UNEX could be utilized from researchers and industrial technology developers within energy micro resources and intelligent control systems.

Given the need for continuous new or refurbished city infrastructure, a parallel ecological vision of new projects, can seriously raise issues of changes and alterations that these projects may bring to the natural environment. Via this platform new city projects can be assessed on the disturbance and changes in relation to the balanced physical connectivity. For instance, alteration of hydrological processes of areas susceptible to these urban projects is a critical issue. Such infrastructures could impact in a negative way on ecological processes, with a significant impact on food chains and environmental risks of those areas. Exergy could be utilized to measure the environmental impact as well defining sustainability and viability of all related parameters.

6.2.1 Deep Learning and Energy Resource Management

A scientific field which remains untapped even though a large amount of manpower is devoted to its exploration is Artificial intelligence (AI) and big data. The amount of available and unexplored data is huge and beyond anything we could have thought so far. A good example is the automotive industry. Until recently, all information stored in the car's "brain" was lost every time we were removing our key from the engine. Now things has changed. Information regarding range, road inclination, engine status, tires situation, gasoline or battery state, nearest route, closest gas stations, and millions of data (big data) are saved and could be used later or during driving. Exactly the same thing could happen within cities. An observer could store and use climatic information, incidents, traffic jams, buildings status, wind speeds, and an unbelievable number of data that now we cannot even imagine are useful data.

In many countries or, at times, at the city level there are declared goals of reaching a 100% fossil-free overall energy mix by the years 2040 or 2050, with various milestones such as 100% Renewable Energy Sources (RES) in the heat and/or electricity and/or transport sectors etc. The importance of keeping momentum in the investment ecosystem is critical [Seljom and Tomasgard, 2015]. Even though several countries – at least in Northern Europe have relatively large capacities on cross-border energy trading and balancing to and from other neighboring countries the multi-faceted interconnection issue is starting to challenge the future prospects of the ambitious energy plans. Moreover, due to convincing and dynamic national plans in a number of renewable energy projects, the interconnectors between neighboring countries has been severely limited in recent years, due to congestions and co-generation. This results many times in curtailment, a large number of hours of negative prices every year, and wasted energy. The need for finding a more concrete solution for a more flexible system that will incorporate various loads and minimize energy wasted – especially within cities – is huge.

Via the Urban Nexus Platform and AI this is possible. One option is plant factories. But how can plant factories be related to renewable energy sources? Indeed, can a small Scale Urban Plant Factory with Automated Lighting (ssPFAL) be a viable investing option within city centers in order to promote food supply security, in the ever-expanding smart city environment, and at the same time energy balance energy production and demand? Can all unused urban information be collected via AI and utilized in some way to benefit the energy demand generation scheme, and in what way?

The approach behind this statement is to first find and register when power in urban and suburban areas is needed, coupled with information derived via extensive AI applications develop and investigate the effect of large deployment of plant factories (hydroponic units) into the power systems. Since most of the developed and developing countries include a high share of renewables and are interconnected to several markets – which is where most European countries want to reach, this idea can become a worldwide successful paradigm for sustainable cities future design. The hypothesis is that by decreasing their power supply needs, allowing the transmission system operator not to introduce expensive and less clean thermal power plants into the system, the peak demand can be covered and meet the demand by postponing it via the Demand Response (DR) options, especially since plants have a different routines than humans. That could mean, for example, that the plants inside the plant factories could use the artificial light when the electricity price is low (or even negative) and the required darkness when the price is high. That means – under a large deployment plant within a city – that higher amounts of renewable energy could be absorbed/integrated and local vegetables production be increased, which subsequently will mean lower CO_2 emissions since there won't be any need for leafy products of the city of Seattle to require "imports" from California. Therefore, exergetically-wise the losses will be expected to be lower than in any other case and at the same time the environmental impact a lot smaller.

This idea, will improve the performance of smart cities platforms and the power system operation, which shall lead to increased efficiencies of the system and reduced operational costs. It will assist city planners and decision makers to plan new green investments and at the same time monitor smart cities operation via KPIs and other assessment criteria. And this could happen only by studying the electricity prices and the performance of the plants within a plant factory. Imagine if all the untapped resource of big data within cities is utilized what could happen!

6.2.2 Beyond Conventional Approaches in the Built Environment – How Can We Specifically Move a Step Further?

More than half of the world's population is living in urban areas [United Nations Human Settlements Programme, 2016]. In contrast to the current situation, where cities run as inherently inefficient living organisms, AI and big data can transform our cities in self-sustained factories that could even produce energy/water/food. The concept of energy positive neighborhoods is well known, however with very little attention gained and unclear positive outcome for the urban population. The goal is to work with the challenges related to energy within cities and focus on the development of UNEX, the platform aiming at improving the exergetic efficiency of the cities. The societal approach should be on working closely with scientific and industrial international partners within the fields of smart grids, sustainable buildings and AI/deep learning. A number of unexploited natural phenomena that take place within cities, not being translated into data and information, should start being studied, for instance climatic phenomena by using Computational Fluid Dynamics (CFD) analyses, deep learning etc on the basis of achieving higher renewable integration and healthier living environments in the urban and suburban environment.

6.2.3 Urban Climatic Phenomena and Detailed Analyses

Computer aided simulations are already playing a huge role in understanding the world around us. Computational Fluid Dynamics simulations are utilized when trying to understand the flow of liquids within a well-defined environment. A specific example could be the wind flow within city centers. Various turbulence models such as RANS, LES as well as a number of hybrid and novel methods could be implemented throughout. Detailed models of various cities, as case studies, could be designed and printed in a 3D printer. Such models can be used in order to validate the CFD models under investigation through conduction of experiments within a wind tunnel for instance. The CFD models under investigation, coupled with mass balance, temperature and buoyancy models will corroborate the sustainability and economic viability of different man-made interferences in the present state of the built environment. The positive or negative sign applied to such an evaluation model, will pinpoint the direction to which policy makers, engineers and markets should strive in order to promote ideas for a more viable future. The expected outcomes of such an approach could be identifying the areas within the city that are expected to experience high heat losses due to high wind speeds there. This way, these sensitive areas will be revealed. At times, when high wind speeds are accompanied with low quality or no building insulation could result in areas where specific measures should be taken. Mapping the spots within cities with low wind speeds could also mean that we have identified potential sites for mass ssPFAL deployment (since a specific temperature should be maintained to keep herbs, fruits, and vegetables' growth rate high) and mapping the spots within cities with high wind speeds could potentially be translated into available areas for new RES projects.

The overall approach is that the ssPFALs could act as flexible loads under specific DR consumer programs. Unused areas, terraces, open spaces, could be explored and selected in order that the ssPFAL be installed.

The development of ssPFALs inside the urban environment and the set of short food supply chains and networks will act beneficially for local communities by means of job creation, CO_2 production decrease, efficient load management, strengthening of the economy and greening of the urban environment (social impact analyses are needed).

6.2.4 AI, Deep Learning and Unused Information for Various Applications

However, this is simply the tip of the iceberg regarding big data within cities. The unexplored information is a unique resource that has not been explored by the scientific society. The data we referred to earlier were based simply on ideas and not on specific methodologies or methodological frameworks already established. This is happening simply because the exploitation rate is and will remain too low till we understand the potential of this scientific sector.

Deep learning could be utilized to interpret simple urban phenomena, such as trees swaying. It could translate trees swaying into wind speeds and therefore identify good spots for urban wind turbines. Via wi-fi trail cameras, videos/images could be collected and trained via visual software tools such as lobe and AWS DeepLens (Fig. 6.3), which shall build custom deep learning models and begin training. Then, the trained model could be sent to CoreML or TensorFlow and run it directly in an app producing real-time big data that can be utilized for automated decision making algorithms for smarter usage of heat pumps, or more (or less) flexible loads in or out of the system (such as plant factories), or higher wind integration (via smarter planning for new wind energy projects).

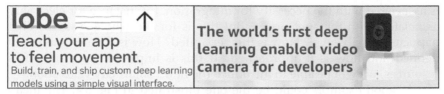

Fig. 6.3. Lobe and AWS DeepLens AI platforms

6.3 A Bi-directional Exergetic Efficiency Approach

Among the various tools that deal with environment management and energy resources and have been applied individually to specific processes or products are Exergy-based assessments. We should further develop and integrate Exergy-based methods (and Life Cycle Assessment (LCA)) onto the built environment. The application of specific tools to the built environment in order to increase efficiency and reduce environmental impact will intensify interdisciplinary collaboration and methodological sharing/comparison within big cities.

Decision makers and urban planners take into account three points that are innovative in the exergy-based assessment approach.

1. Different energy forms have different qualities, their exergy value but also their related emissions during processing and use. Their use depends on the characteristics and the necessities they have to satisfy within the city.
2. A clear distinction should always be made among various energy carriers (electricity, fuels) and primary energy sources (coal mines, oil reserves, renewable energy sources) for evaluating the overall performance of energy systems. The same useful energy supplied to a city's part could come from different amounts and sources of primary energy and is then associated with different environmental impacts.
3. The concept of exergy makes it possible to express in a quantifiable form the technical aspects of sustainability.

What has been neglected so far is the demand side. So far we have been estimating the exergy values as net numbers and according to those evaluating a process as more exergetically efficient or less. A step forward was when we introduced the **Exergetic Capacity Factor (*ExCF*)** of a power plant [Xydis, 2013b; Petrakopoulou et al., 2017; Midilli et al., 2016; Dragan et al., 2011; Fonseca et al., 2007; Sun et al., 2012] where we included all the unconsidered till then losses. However, this was still far from the dynamic estimation of Bidirectional eXergetic efficiency or **Bidirectional Exergetic Capacity Factor (*BiXef*)** that is introduced now. *BiXef* is linked to consumption (or integration) either as a real-time index or as an overall index of resource usage depending on the demand. It is clear that *BiXef* is associated with the necessity for integration, for instance for electricity associated with the spot prices. But this is for the whole system. What happens at a city level? How is this separated? How is different necessity defined and how is it approached? If this is implemented at all city's resources, we will be able to introduce the new "smart city v2.0" soon.

A good example could be the one illustrated in Fig. 6.4. The mean hourly leafy products demand variation is illustrated with the yellow columns. But the production and the available quantity within a specific neighborhood at 8.00 am is where the star is and at 15.00 where the little text cloud is. That means that in the morning the demand can be met, while at 15.00 it cannot.

Following the approach of [Xydis, 2013b], the Exergetic Capacity Factor (*ExCF*) of a proposed wind farm, including all losses can be estimated as:

$$ExCF = \frac{NetAEP}{8760 \cdot Ci} \cdot 100\%$$

where *NetAEP* is the Net electricity generated, *8760* is the number of total hours throughout a year), and *Ci* the nominal capacity of the wind farm. *NetAEP* is the net yearly electricity produced independently whether this production was integrated or not and if it was integrated when it was greatly needed. Via *BiXef*, this is going to be taken into account. Bidirectional *ExCF* could be calculated by the following formula:

$$BiXef = \frac{NetAEP}{8760 \cdot Ci} \cdot \frac{\left|RTP_{ti} - RTP_{ti-1}\right|}{RTP_{avs}} 100\%$$

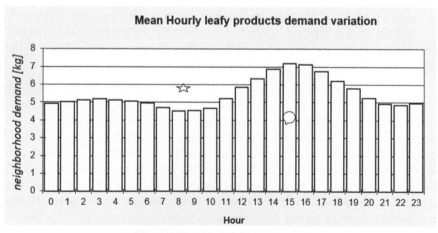

Fig. 6.4. Mean hourly leafy products demand variation

Where RTP_{ti} is the Real Time Price at hour t_i and RTP_{ti-1} at the previous hour and RTP_{ave} the average Real Time Price throughout the year. This means that the second part or the equation could be a value greater than 1 which could indicate the need for energy integration into the system. Especially in cases where there is a high real time price and suddenly the next hour there is negative price. This where the system requires immediate response such as in the example of Fig. 6.5.

The same analogy could be utilized in all other products, such as the need for herbs, or the water supply based on the demand and the production. In the nexus we are going to see in Chapter 7 this could be of significant importance. This approach could minimize the system losses and assist towards a more efficient and integrated smart city.

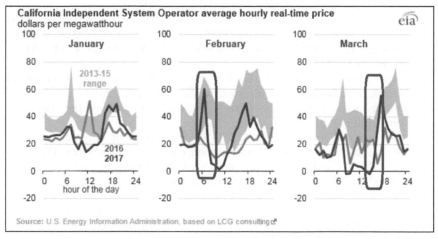

Fig. 6.5. California System Operator average hourly real time price (USD/MWh)

6.4 Concluding Remarks

What is absolutely necessary is to study in depth the efficiencies related to production and demand during the city growth periods under a nexus concept. Exergy analysis on its own could offer constructive results, however, introducing innovative indexes, such as the BiXef, could give a more holistic systematic view of the nexus. Under a nexus platform, using a lot of data and AI a number of applications could be optimized. For instance, when the need for grocery for a neighborhood is random and gives nothing back as knowledge for the whole system. But when all products and all shopping are registered in a unique community platform the overall efficiency of the system could be estimated. This way, by using the BiXef index we could know when local production is enough or not. This could lead to the optimization and improvement of the various commercial subsectors within a city since it provides the real linkage between the scientists and the urban environment.

REFERENCES

Arribas, C.A., Blazquez, C.A. and Lamas, A. (2010). Urban solid waste collection system using mathematical modelling and tools of geographic information systems. Waste Management and Research 28(4): 355–363.

Dombayci, Ö.A., Gölcü, M. and Pancar, Y. (2006). Optimization of insulation thickness for external walls using different energy-sources, 2006. Applied Energy 83(9): 921–928.

Dragan, M., Uzuneanu, K., Panait, T. and Gelu, C. (2011). The exergetic evaluation for the steam boiler. Recent Advances in Fluid Mechanics and Heat and Mass Transfer – Proc. of the 9th IASME/WSEAS Int. Conf. on Fluid Mechanics and Aerodynamics, FMA'11, Proc. of the 9th IASME/WSEAS Int. Conf. HTE'11, pp. 267–270.

European Parliament and the Council of the EU, Directive 2010/31/EU of the European Parliament and of the Council of 19 May 2010 on the energy performance of buildings (recast). Official Journal of the European Communities (2010).

European Union. Directive 2012/27/EU of the European Parliament and of the Council of 25 October 2012 on energy efficiency, amending Directives 2009/125/EC and 2010/30/EU and repealing Directives 2004/8/EC and 2006/32/EC. Brussels: European Commission, 2012.

Fonseca, J.G.S., Asano, H., Fujii, T. and Hirasawa, S. (2007). Study on the effect of a cogeneration system capacity on its CO_2 emissions. Challenges on Power Engineering and Environment – Proceedings of the International Conference on Power Engineering 2007, ICOPE 2007.

Hirano, Y. and Fujita, T. (2016). Simulating the CO_2 reduction caused by decreasing the air conditioning load in an urban area. Energy and Buildings 114: 87–95.

Jim, C.Y. and Tsang, S.W. (2011). Biophysical properties and thermal performance of an intensive green roof. Building and Environment 46(6): 1263–1274.

Koroneos, C.J., Nanaki, E.A. and Xydis, G.A. (2012). Sustainability indicators for the use of resources—The exergy approach. Sustainability 4(8): 1867–1878.

Matic, D., Roset Calzada, J. and Todorovic, M.S. (2016). Renewable energy sources—Integrated refurbishment approach for low-rise residential prefabricated building in Belgrade, Serbia. Indoor and Built Environment 25(7): 1016–1023.

Midilli, A., Inac, S. and Ozsaban, M. (2017). Exergetic sustainability indicators for a high pressure hydrogen production and storage system. International Journal of Hydrogen Energy 42(33): 21379–21391.

Nanaki, E.A., Koroneos, C.J., Roset, J., Heidrich, O., López-Jiménez, P.A., Susca, T., Christensen, T.H., De Gregorio Hurtado, S., Rybka, A., Kopitovic, J., Heidrich, O. and López-Jiménez, P.A. (2017). Environmental assessment of 9 European public bus transportation systems. Sustainable Cities and Society 28: 42–52.

Perini, K., Ottelé, M., Fraaij, A.L.A., Haas, E.M. and Raiteri, R. (2011). Vertical greening systems and the effect on air flow and temperature on the building envelope. Building and Environment 46(11): 2287–2294.

Petrakopoulou, F., Robinson, A. and Loizidou, M. (2016). Exergetic analysis and dynamic simulation of a solar-wind power plant with electricity storage and hydrogen generation. Journal of Cleaner Production 113(1): 450–458.

Seljom, P. and Tomasgard, A. (2015). Short-term uncertainty in long-term energy system models—A case study of wind power in Denmark. Energy Economics 49: 157–167.

Sun, F., Fu, L., Zhang, S. and Sun, J. (2012). New waste heat district heating system with combined heat and power based on absorption heat exchange cycle in China, 2012. Applied Thermal Engineering 37: 136–144.

Susca, T., Gaffin, S.R. and Dell'Osso, G.R. (2011). Positive effects of vegetation: Urban heat island and green roofs. Environmental Pollution 159(8–9): 2119–2126.

Susca, T. (2012). Multiscale approach to life cycle assessment: Evaluation of the effect of an increase in New York City's Rooftop Albedo on human health. Journal of Industrial Ecology 16(6): 951–962.

United Nations Human Settlements Programme, World Health Organization, Kobe Centre (2016). Global report on urban health: Equitable, healthier cities for sustainable development. WHO Kobe Centre, Kobe, Japan. Available at: http://apps.who.int/iris/bitstream/10665/204715/1/9789241565271_eng.pdf

Vairavamoorthy, K., Gorantiwar, S.D. and Pathirana, A. (2008). Managing urban water supplies in developing countries – Climate change and water scarcity scenarios. Physics and Chemistry of the Earth 33(5): 330–339.

Vardopoulos, I. (2018). Multi-criteria decision-making approach for the sustainable autonomous energy generation through renewable sources. Studying Zakynthos Island in Greece 7(1), doi:10.5296 /emsd.v7i1.12110

Xydis, G. (2012a). Effects of air psychrometrics on the exergetic efficiency of a wind farm at a coastal mountainous site – An experimental study. Energy 37(1): 632–638.

Xydis, G. (2012b). Wind-direction analysis in coastal mountainous sites: An experimental study within the Gulf of Corinth, Greece. Energy Conversion and Management 64: 157–169.

Xydis, G. (2013a). Wind energy to thermal and cold storage – A systems approach. Energy and Buildings 56: 41–47.

Xydis, G. (2013b). The wind chill temperature effect on a large-scale PV plant – An exergy approach. Progress in Photovoltaics: Research and Applications 21(8): 1611–1624.

Xydis, G.A., Nanaki, E.A. and Koroneos, C.J. (2013). Low-enthalpy geothermal resources for electricity production: A demand-side management study for intelligent communities. Energy Policy 62: 118–123.

Zinzi, M. and Agnoli, S. (2012). Cool and green roofs: An energy and comfort comparison between passive cooling and mitigation urban heat island techniques for residential buildings in the Mediterranean region. Energy and Buildings 55: 66–76.

Spatial Planning and Exergy – Design and Optimization

G. Xydis[1] and E. Nanaki[2]

[1] Department of Business Development and Technology, Aarhus University, Birk Centerpark 15, 7400 Herning, Denmark
[2] University of Western Macedonia, Department of Mechanical Engineering, Bakola & Sialvera, Kozani 50100, Greece

7.1 Introduction

GIS consist of an integrated technological super-system, combining software, hardware, data, and the human factor. It is not easy to define GIS, because of the variety and the complexity of their fields and the blurry centralization of their focus. The diversity of GIS applications are especially useful for agriculture, botany, mathematics, economics, computing, photogrammetry, surveying, zoology, and geography.

A system for capturing, storing, analyzing, and displaying data which are spatially referenced to the Earth was the most common definition utilized. Others relate the usage of the "database" and "decision support system" to the appropriate GIS definition, linking it to case-by-case demands and needs.

7.1.1 How GIS-based Research Should Start and What should be Included

Goodchild (1987) points spatial analysis to the "set of analytical methods which require access to both the attributes of the objects under study and to their locational information". Figure 7.1 presents a typical example of GIS analysis (a 3D mean wind speed map) in a Greek mountainous area.

The computer-aided simulations for project engineers/decision makers via tools such as QGIS or ArcGIS etc are used often to produce

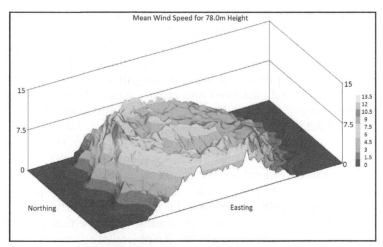

Fig. 7.1. 3D mean wind speed map
Color version at the end of the book

an outcome of a potential image (map) on the most suitable areas for a project in accordance with the legal restrictions related to distance limits etc. In general, it seems that GIS constitutes individual sub-units, which are essential for the creation of a geographical information set. A system like that could answer, through its applications, different queries such as location, condition, routing, pattern, modeling, which can provide the required results. Some of the key features that differentiate GIS from other information systems are the general focus on spatial relationships, together with specific attention to analytical and geographically-focused modeling operations. It is the ability to organize and integrate disparate data sets together using i.e. spatial searching and overlay operations that make GIS so powerful. It is common for GIS to be used as a "Decision Support System", i.e. a tool which is able to elaborate a huge number of geospatial data through a variety of methods and techniques [Noorollahi et al., 2016].

In several projects the methodology being used when dealing follows a similar framework. What it is usually known is the policy-based and law-enforced restrictions about the distances of different categories, such as environmental, geographic for residential areas, highways, airports etc. Therefore, what we can know about this project lies on realism because this is a numerical approach and the potential outcome each time is totally objective. The way we usually produce results is through a computer-aided process, where maps for a variety of spatial categories were mapped and numerical data about the restrictions are added to the maps, in order to illustrate the focus sites. The first steps in the representation of potential project locations are 1) data collection and 2) the creation of a spatial database from which the relevant maps could be extracted. After the data collection, a geo-referencing approach follows, where the spatial database was created and the selection of the right attributes were separated and again clustered in order to create the desired layers. Therefore, all the geo-references are implemented into the computer program (e.g. QGIS), in order to be assessed, analyzed and create maps. These maps usually illustrate the buffers, which means the distance limits that the legislation provides. The detailed flow diagram is depicted in Fig. 7.2.

There are multiple perspectives we need to cover through a GIS study. A geographic model usually includes plenty of variables, which we need to take into consideration like physical, environmental and human impact factors on any project site suitability investigation [Rodman and Meentemeyer, 2006; Panagiotidou et al., 2016]. The results are usually either just maps or roadmaps aiming to prove that projects are "technically and economically feasible". An objective approach allows the user to extract certain results as the outcome. The critique of an objective approach is that there is wide variety in data collection and certainly results in the outcome, as long as the research depends on results from computer program calculations on various case studies of a geographical approach and using a specific methodology (Fig. 7.3) in order to optimize projects' location.

In order to assess the results of every study we need to validate the data. Reliability always involves demonstrating that the operations of a study, such as data collection procedures, can be repeated with the same results.

The evaluation should always be part of the research output – held though validation of the outcome – a solid chain of evidence built by multiple sources of data selection, so the evaluation follows construct validity strategies. Furthermore, pattern matching often used in

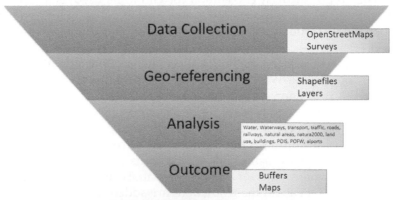

Fig. 7.2. Generic representation of working with GIS tools
Color version at the end of the book

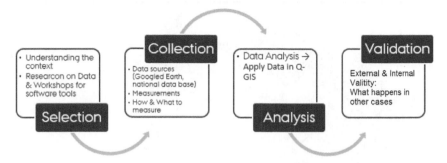

Fig. 7.3. Methodology followed in a common GIS-based research project

projects, add internal validity in those and the theory-based approach in combination with replication logic confirms the external validity.

Every GIS-based research purpose is the digitalization of previous studies, presentation within a GIS system, and a creation of a functional database. The most significant benefit deriving from the completion of such studies, is the ability to use geostatistical methods in the focus areas. Researchers – after the in depth analysis and digitalization – can use this geodatabase for several uses over the years in the future.

Raw data were gathered from the digitalization of the study and at timesdataset are freely distributed through official sources (e.g. each countries GIS repository) via digital services. Road networks, rivers, tourist attractions, Natura 2000 sites, land cover dataset, settlements' limits, and mining zones geodatabases could freely be downloaded through governments' official geo-repositories (Fig. 7.4).

Furthermore, quite recently, the Google Dataset Search (beta version) was launched providing the unique opportunity to researchers/city planners etc to acquire useful planning, environmental or other online data that could be downloaded and used immediately (Fig. 7.5). The product is targeted at data journalists, scientists governmental employees, organizations and by using structured data formats when uploading data on the Internet, such as schema.org, the online data community will grow and more analyses shall lead to more accurate results in all scientific fields.

7.1.2 Techniques Used for Data Processing; Problems Encountered

Once raw data are acquired from different sources, check, rating and classification usually follows. Data recorded and published are the basis for those researchers that digitize. Possible errors could be found, Inexperience on digitizing the raw data or even marks of negligence in certain cases. Technical problems can mainly occur because of the complexity and the size of the data and at the same time the low-ability

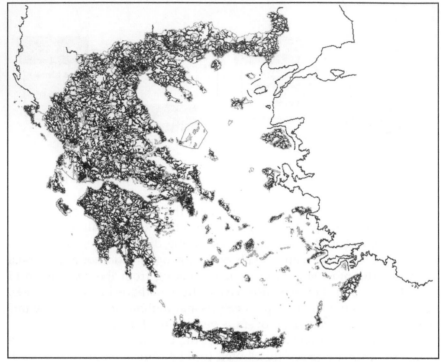

Fig. 7.4. GIS data presented all in one layer for Greece
Color version at the end of the book

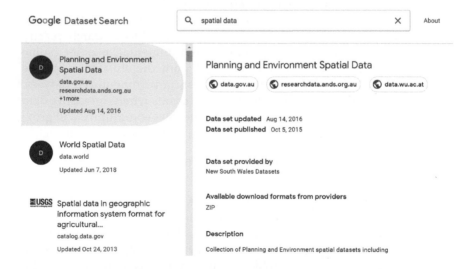

Fig. 7.5. Google Dataset Search

computer used in cases to process the data, even projection of the data (different coordinate systems), to create the maps and to run the statistics. Researchers usually operate with ArcGIS, SPSS, surfer, and Microsoft Excel tools to process the data collected. In most of the cases in the field, data are outdated over a period of decades, but research moves forward. Updated tools are out, data resolution is improved (for instance there are cases where Digital Elevation Models can be found for a less than 1 m of accuracy – which a decade ago was a utopia. The best analysis one could use was 90 m SRTM data). Now, data mining, data management, artificial intelligence and spatial analysis tools are used in a combined form to produce the classification maps and get the statistical results that would help us confirm the original hypothesis each time.

The selection of the data is always done via desk research for data on the Internet and access to sources. For every case, data in the form of shapefiles are downloaded databases linked to other databases such as OpenStreetMap. We, as researchers, should always aim at daily up-to-date data, which means that this source of information is valid and timely updated.

OpenStreetMap is a source of data, which refers to streets, shops, transport stations, and more, a team of people, with expertise on geographical information and cartography. They have created this base of map data, where everyone can have access and use whatever it is needed, as long as they cite the project properly.

Shapefiles are a widespread data format with geographic and geomorphological data designed to be used, studied and processed by all (most common) GIS programs. These files were initially developed as an open specification option for data interoperability between Esri and other GIS software products. The shapefile format can describe spatial vector features such as points, line and polygons, representing, for example, water wells, rivers, mountains and lakes. Each element typically has features that describe it, such as name or temperature.

One of the most common programs utilized, the QGIS program, is a free and open source geographic information tool, which creates, edits, visualizes and publishes geospatial information, available for any user and any operation system. There is a variety of documentation, guides, tutorials and manuals online, so anyone with basic software understanding ability can easily learn how to use QGIS, in order to elaborate spatial data.

7.2 Optimization via RES Deployment within Cities

Spatial planning is widely used within cities. Sustainable Urban Planning and Smart Cities both have evolved from subfields into dynamic and independent research areas.

Developed countries are underway to reaching their national goal of high percentages of renewable energy in the electricity production mix. With declared goals of reaching a 100% fossil-free overall energy mix by some countries in the 2050 (e.g. Denmark) a bit later for some other countries, with various milestones in the meantime such as 100% EVs [Lévay et al., 2017] or of 100% RES in the heat, electricity and transport sectors by 2035, the importance of keeping momentum in the investment ecosystem is critical [Seljom and Tomasgard, 2015]. There are countries with strong links to neighboring countries regarding electricity interconnections, however as the RES shares are increasing, negative prices and curtailments – which means losses – are increasing. For instance, Denmark has relatively large capacities on cross-border energy trading and balancing to and from the neighboring countries Germany, Sweden and Norway, the multi-faceted interconnection issue is starting to challenge the future prospects of the ambitious energy plans. As a result of the significant increase in numbers of wind turbines installed in northern Germany, the interconnectors between Denmark and Germany has been severely limited in recent years, due to congestions and co-generation. This results several times in curtailment and wasted energy. Curtailments and down-regulation shall continue to occur, as the wind energy share shall increase. This also puts investments in Denmark under pressure. Since there are cases in central Denmark where wind farms are shut 30% of the expected operating time, investors and stakeholders are urgently on the look out for a viable solution. In 2016, the average wholesale price of electricity generated by wind turbines was down at 0.025 EUR/kWh, which is more than 10% below the average electricity price on the Danish wholesale market the same year. This means, in practice, that during the last few years, investors have no profit.

The above mentioned case is becoming more and more profound in cities and urban areas as the need for distributed renewable energy sources integration and participation is needed in order all the near cities' thermal and coal/oil/lignite based facilities to be eventually replaced by green forms of energy. Global carbon dioxide (CO_2) emissions reached an all-time high and despite the measures cumulatively continue to rise in the last years [Koroneos et al., 2012a].

Optimization and replacement of conventional sources for electricity production can be found in several studies. Cases in small cities in isiolated areas/islands are found in literature. Koroneos et al. [2012b] were focused in a small case study for Lemnos Island. A similar case study in Crete Island proved that in suburban areas cold storage assist in optimizing wind integration [Xydis, 2012]. Ji et al. [2016] worked on energy load superposition and spatial optimization in an urban environment, proving that a P2P system among buildings is possible focusing on load curves from different buildings in order that the total energy peak is smaller than the sum of individual buildings' energy peaks. Ge and Kremers [2016] applied an agent based modeling in the context of urban energy planning among heterogeneous systems. Bachmaier et al. [2016] goes beyond that and introduces district heating and energy storage systems mainly for grid balancing, while Alhamwi et al. [2017] introduced various flexibilization technologies within cities using primarily the OpenStreetMap database. Sarralde et al. [2015] focused in increasing solar derived electricity production in London, Xydis [2013] used a model that could increase photovoltaic production when sitting in areas with significant wind speed, while Vermeulen et al. [2018] developed urban models for optimization of passive solar irradiation.

However, although there are numerous studies that are based on optimization within cities, there only few that select exergy analysis as the optimization guide. Via exergy analysis accurate urban flows could be studied (in-depth) and provide unique results that could guide decision makers. For instance, structured nexus or nexi could play a role in the future on urban systems. Via exergy analysis it will be proven that the more nexi we introduce in the city, the lower inefficiencies will be observed. For instance, if we are focused on the food-energy nexus, we will see that it aims to locally produce vegetables more effeciently using less power.

7.2.1 The Food-energy Nexus

The continuing increase of the human population, the competition for land, water, energy supply security and overexploitation of natural resources has led to many changes in the agriculture domain. Urbanizations and industrialization as long as global warming and deterioration of the environment and the natural resources are bound to minimize the available arable land for cultivation and its productivity [Sardare and Admane, 2013]. It is clear that by developing and growing innovative and optimizing materials and by the utilization of the energy resources a better exploitation of soil area for plant growth can be provided. Under these circumstances, providing sufficient yield while meeting the consumers' needs in terms of quality for the entire population using conventional agricultural production methods will become even more demanding [D'Autilia and D'Ambrosi, 2015; Putra and Yuliando, 2015].

A significant number of researchers have concluded that hydroponic greenhouses in the urban environment are an optimal and modern way of cultivation based on the energy supplies, which leads to best correlation between techno-economical evaluation, products and the environment [Liaros et al., 2016].

Coupling of indoor hydroponic systems and relative technologies and mass deployment may lead potentially to significant benefits for the environment and the economy. Indoor hydroponic farming requires

Fig. 7.6. A hydroponic unit, part of the food-energy nexus

significant amounts of energy (heating demand of the plants and lighting) operating 24/7. By controlling the energy requirements of the system, while at the same time optimizing the plants' growth, has great potential leading to the integrated energy-food nexus solution [Xydis et al., 2017]. When electricity prices are high, indoor farming demand could be lowered and electricity prices are low, indoor farming electricity and heating requirements can be sufficiently met. A mass deployment of such systems, can even help make the grid more efficient by providing the utility Demand Response (DR), so when power is needed elsewhere on the grid, hydroponics can reduce the amount of energy they are using, allowing the utility not to switch on expensive and "dirty" peak power plants. One of the aims of this work is to experimentally identify for each plant the optimal environmental conditions in an isolated soil less environment and define each plant's growth inertia when there is need for switching off or constraining the required energy usage, something which has never been studied so far.

An innovative method of herbal plants cultivation has recently gained the attention of the scientific community. Two of the most renowned supporters of the idea of indoor farming under controlled environment, are Kozai [2013; 2016], Kozai et al. [2015], and Despommier [2012], strenuously support the idea that plant factories or vertical farms with their indigenous characteristics might provide the solution to sustainable food supply to the urban environment. These plant factories will lead to the creation of new green urban regions, will offer food and raw material security to interurban regions which suffer from lack of culture soil. In addition, the proposed concept shall minimize the use of overfertilization in the cultivations and overutilization of the soil deposits. Furthermore, new job positions will be developed in the urban environment. Moreover, the reuse of the abandoned buildings or local factories will be achievable and will result to the minimization of the cost for storage and transportation. CO_2 emissions induced by transport activities related to vegetable distribution shall be reduced significantly annually .

7.2.2 The Transportation-energy Nexus

Similarly, alternative small-scale transportation options within cities could improve drastically the inhabitants' quality of life. During the last decade the electric vehicle industry has been developed rapidly due to the associated environmental impacts arising from the use of conventional fossil fuel-based Internal Combustion Engines (ICE). This turn to sustainable mobility presents several positive effects for the environment including air quality improvement, noise reduction, and fuel independency in the case of renewable sources (RES) utilization. Specifically, in the European Union, the transportation sector comprises more than 33% of the EU-28 final energy consumption, and in 2014 was responsible for the annual emissions of at least 1147Mtn CO_2 equivalent (i.e. 30% of the CO_2 EU total emissions). This suggests that by improving efficiency in the transportation sector, or by turning to zero-carbon emissions vehicles, environmental impacts associated with transportation will be minimized [Union Européenne and Commission Européenne, 2015].

However, most electric vehicles nowadays present drawbacks concerning their establishment as a reliable transportation medium. These include limited driving range, long recharging time, deep discharging problems of the battery bank, and high cost [Serrao et al., 2009; Helmolt and Eberle, 2007]. To this end, research and development (R&D) of hydrogen based technologies indicates their high potential in the transportation sector. Hydrogen presents several intrinsic characteristics that make it attractive as an energy carrier for transport applications. It can be produced in large quantities from a range of primary sources, though the most interesting is the production from RES. A clear advantage compared to batteries is that hydrogen can be refuelled just as gasoline in a few minutes [Martin et al., 2009].

Nowadays, hydrogen can be used in a large range of applications, where fossil fuels dominate. Nevertheless, the fact that H_2 does not exist in its molecular form in nature makes its effective production important. Although there are several available methods for producing hydrogen from RES, the most promising one comprises water electrolysis. Electrolytic hydrogen presents significant advantages over other widely spread methods of energy storage such as batteries as it allows storage of large energy amounts, up to GWh, for a long period of time. Furthermore, a very promising technology is the SOEC which presents high efficiencies but the high operational temperature above 800°C, limits its range of applications, in general.

PEM electrolysis, on the other hand, is nowadays considered the best solution regarding a large variety of applications. Its operating principle is based on the use of a polymer electrolyte which is pressed between two electrodes, and immersed into pure water of low conductivity. The anode catalyst is usually Iridium (Ir) while Platinum (Pt) or Lead (Pb) is used in the cathode. Through the application of Direct Current (DC), oxygen evolution takes place at the anode. The hydrogen ions are transported to cathode across the polymer electrolyte where hydrogen is generated. The main advantages of PEM electrolysis are production of high purity gas, high efficiency, compact design, and safer operation due to the absence of liquid electrolyte [Millet and Grigoriev, 2013]. A hydrogen system also includes the storage process which can be accomplished with three methods: As a gas in compressed cylinders, as liquid in special low temperature tanks, and as solid stored in metal hydride canisters.

The main disadvantage of gaseous storage is the large required volume of the tank. However, at this time it is considered the most cost effective method particularly for small scale applications. On the other hand liquid storage of H_2 presents significantly high demand of energy for reserving hydrogen at cryogenic temperatures. Metal hydrides is the most promising technology which requires low volume and energy demand, but the canister's large weight and cost confines its range of applications [Riis and Sandrock, 2006].

It was mentioned that production of H_2 via RES consists a very promising method for portable applications. Wind energy exhibits a stochastic and variable availability, enhancing the mitigation of the RES maximum penetration during the daily and seasonal electricity demand fluctuations. Therefore, even in the case of high wind potential, the produced energy may not be integrated into the electrical network, resulting in a waste of energy and monetary losses for RES investors. Hence, one may notice that it would be beneficial to combine the potential of hydrogen mobility with the wind energy curtailments in order to deploy a new market including hydrogen powered vehicles and production of hydrogen from the otherwise curtailed power from wind farms. This way, within cities the losses again – exactly as in the food-nexus nexus – will be lower.

7.2.3 Urbanization, CFD, and a Proposed Experimental Approach

- Urbanization refers to the process of populace shift from rural to urban areas. According to United Nations Population Fund [2016], urbanization is closely linked to a vast range of disciplines such as politics and sociology, economics and urban planning as well as public health. Loss of greenery and urban densification is associated with mental health [Haaland and van den Bosch, 2015], perceived general health and mortality [van den Berg et al., 2015]. To be more specific, urban air quality in Europe, often surpasses the legislation standards, increasing by this way the environmental risk and posing the greatest threats to public health [Guerreiro et al., 2015].

Currently, more than half of the world's population lives in urban areas [United Nations Human Settlements Programme, 2016]. In 1960 urban population accounted for just above 34% of the global population. In 2008 that figure surpassed 50% [The World Bank, 2018] while forecasts predict that more than 70% of the humanity will be an urban citizen by 2050 [UNDESA, 2014].

Cities, in order to support their habitats, require large amounts of resources. From a physical perspective, the urban environment acts as a huge source and sink of energy and matter. The urban environment, due to its expansive nature, needs to be attractive in the eyes of potential residents. Warranting high quality of life standards, necessitate a multitude of services such as healthcare and education, culture and safety [Phillis et al., 2017].

The densification of urban areas, raises concerns regarding their regional sustainability and the role those play in the global energy and matter flow and balance. In this context it is imperative to produce resource efficient centers that will circumvent biodiversity degradation, hindering of the environment and social qualities [Power, 2001].

The resilience and sustainability of human settlements is considered to be of major importance and is marked as one of the sustainable development goals by the United Nations. The anticipated swift expansion and densification of the built-up environment, as regards the urbanization of rural settlements, directed the interest of researchers towards defining and weighting those characteristics that are most important for the development and promotion of sustainable habitats [Martos, 2016].

The proposed project, having realized the economic and environmental, health and social effects of urbanization as an effect, will incorporate research elements such as literature review, simulation and experimentation in order to gauge the microclimate of urban areas in terms of exergetic, environmental and economic factors ending up to a Food-Energy Nexus (FEN). Furthermore, it is in the intents of the researcher to encourage the growth of novel, small to medium scale business incentives that will promote the sustainability of urban settlements.

Urbanization is an anthropogenic procedure shaped by means of substituting the natural surroundings by what is commonly referred and implied as urban environment. This process creates a unique microclimate characterized by higher temperatures when compared with their natural counterpart. This effect is commonly known as Urban Heat Island (UHI) and was first mentioned by Manley in 1958 [Gordon, 2006].

UHI should not simplistically be interpreted as a spatial temperature increase inside the urban environment. As a phenomenon, it is related to health implications alongwith energy utilization. For example, different types of buildings and their localization inside the same city may be differently influenced when dealing with buildings' matter, environment, and energy footprint. Here it has to be noted that those effects may not be necessarily negative [Davies et al., 2008; Xydis, 2012; Hirano and Fujita, 2012].

Deconstruction of urban microclimate's dependency on key physical parameters is important in order for urban designers, architects and engineers for planning and building sustainable and safe settlements [Mills, 2006].

According to Mirzaei and Haghighat [2010] urban microclimate studies are crudely categorized as either observational or simulations. More recently, the increase in computational resources has encouraged the application of more simulation studies. Furthermore, simulations are less time consuming while not affected by spatial limitations [Moonen et al., 2012].

Currently, there are two market dominating simulation approaches, namely Energy Balance Modeling (EBM) and Computational Fluid Dynamics (CFD). The latter, while more demanding in terms of computational power, when considering urban climate are preferable. CFD may perform simulations in finer scales while coupling velocity and temperature fields with humidity and pollution ones, areas where EBM comes to a halt. With CFD simulations may be employed for studying the urban microclimate in spatial scales ranging from the meteorological mesoscale to the indoor environment [Mirzaei and Haghightat, 2010; Murakami et al., 1999].

Toparlar et al. [2017] through an extensive review of CFD analysis of the urban microclimate concluded that future CFD research should be:

(a) subjected to detailed validation through experimentation,
(b) expanded to engulf case studies of cities located in the developing regions of the world,
(c) focused in Large Eddy Simulation (LES) methods instead of Reynolds Average Navier – Stokes (RANS), since the former is documented to be potentially more accurate in predicting flow fields,
(d) incorporating target parameters such as economy, matter and energy flows,
(e) promoting the use of small scale urban plant factories with automated lighting (ssPFAL) to urban designers by integrating economic factors, investing opportunities and business activities while acquiring new knowledge,

(f) contributing to the renewables high grid integration requirements in modern grid using Demand Response (DR) programs via large deployment of ssPFAL in the urban environment via this Internet of Farming approach.

The need for a detailed urban-based activities energy analysis is of outmost importance. A detailed mapping and efficiency analysis for all the city-related processes will be needed for all cities to eventually become smart cities. It is crucial for SMEs/research institutes/universities to contribute to the expansion of knowledge, regarding the complex dynamics governing the microclimate of densely populated areas, ensuring wind comfort in the respect of their ever-expanding nature. The CFD companies/institutes could experiment in wind tunnels (**Fig. 7.7**) with various turbulence models such as RANS, LES alongwith a number of hybrid and novel methods throughout computer aided simulations. Detailed models of various cities, as case studies, could be designed and printed in a 3D printers. Those models can be used in order to validate the CFD models under investigation through conduction of experiments within wind tunnels. The CFD models, coupled with mass balance, temperature and buoyancy models will corroborate the sustainability and economic viability of different man-made interferences in the present state of the built environment by means of exergoenvironmental

Fig. 7.7. The CET, Aarhus University small-scale wind tunnel

and economic evaluation. The positive or negative sign applied to such an evaluation model applied to variant dead states, according to the principles of exergetic analysis of energy systems, will pinpoint the direction to which policy makers, engineers and markets should strive in order to promote ideas for a more viable and safe future within cities. The CFD analysis shall also uncover where the major thermal losses within the urban environment are expected. In modern power systems the grid can efficiently achieve maximization of RES integration by promoting the participation in Demand Response (DR). The idea behind this project is to examine when power is (in urban and suburban areas) needed, flexible loads can decrease their power supply needs, allowing to the utility not to introduce expensive and less clean thermal power plants into the system in order to cover the peak demand but meet the demand by postponing it via the DR option, especially when flexible loads are not directly associated to humans and their needs. The mass deployment of such flexible loads could be studied and economically analyzed by SMEs, urban planners, and researchers. For instance, those flexible loads could be the hydroponic units mentioned earlier. Food growing and human needs do not necessarily go hand-in-hand. The overall output shall be a city-based framework that could take into consideration exergoenvironmental, CFD, load and integration analyses aiming at a clear message for introducing the new "smart cities 2.0" model.

7.2.4 Remote Sensing-based Spatial Planning and its Importance

Since the availability of satellite data is increased based on the new missions (sentinel satellite missions of the Copernicus Program), the advancements of the technologies, and the accuracy of the acquired data a new door to urban data has opened. Project planners/researchers etc should aim at pointing out the benefits and impacts to the society and businesses of using open and large data volumes in the energy sector.

The European Union has recognized the need to move towards a low-carbon economy. One of the main focuses is to develop advanced short-term forecasting systems ranging from every 15 minutes up to daily and/or weekly basis depending on the demand, taking advantage of the near real-time acquisition of satellite images. For instance, solar irradiation in different topographies, modeling of dust dispersion phenomena, and assessment of cloud cover figures are already utilized in order to achieve accurate forecasting. Accurate wind and solar resource assessment is needed for an efficient smart city environment. As Earth Observation (EO), GIS tools, data acquisition and modeling capabilities move forward, innovative research and commercial applications are developed. The employment of advanced data modeling in combination with open data mining, data retrieval and processing tools, can lead to the production of meaningful value chains for deriving reliable, accurate, stable analysis towards the optimization of energy within the urban environment. Temperature and moisture profiles, precipitation and surface properties, wind and solar energy fields, and essential atmospheric and climate variables, are gathered for the production of advanced marketwise product solutions, creating new cross-sectorial business opportunities to European industries and increasing the overall competitiveness of the European SME energy sector.

Existing forecasting tools – neural networks and stochastic optimization mainly – and power planning tools are used together with open Remote Sensing (RS) based climate and atmospheric data, either to correlate and validate results from the current methodology, or to create completely new information products in order to ensure best wind and solar power integration so that the uncertainties can be taken into account in the wind farm and PV park planning and operation worldwide. The goal is to lead to better models and methods for prediction of heat load and electricity generation and demand within cities but also in rural areas.

7.2.5 Mapping Electricity Demand during Heat Waves and Urban Environment

The rise in external ambient temperatures in urban environments, compared to rural environments, is associated with a series of interconnected impacts, namely comfort, electricity usage increase etc, to meet increased comfort requirements. Energy demand for buildings' cooling is an indicator of the impact of climate specifically on the energy sector. Cooling Degree-Days and hours (CDD and CDH) are the most common practical way for assessing the effect of air temperature on the energy performance of a building and are used as a reasonable approximation of the cooling energy needs of a city with respect to it. In recent years the majority of houses and work environments have air-conditioning (or at least air-controlled) systems. In the case of a heat wave event, power demand is increased greatly often causing significant region-wide blackouts. Then population health is also at risk. By definition, a CDH is recorded if the average air temperature for the hour rises above a base temperature that differs between cities. The spatial distribution of CDH in a city is related to the demand for air conditioning with the energy suppliers sometimes finding it difficult to cope with increased power demand. Until recently, CDH could be only calculated at certain points. The lack of adequate number of ground stations prohibits any meaningful spatial analysis at and around the urban web. Satellite derived LST images at 1km spatial resolution offer adequate spatial but sparse temporal sampling. Downscaled LST images at 1km on an hourly basis can be a strong alternative, after adequate correlation is found between base air temperature with LST.

To this end, based on the mapping results, the goal is to focus on the end-user applications improving at the same time renewable energy integration to the grid. Knowing the LST (and assuming that it can be correlated with the air temperature), the flexibility of electricity generation facilities will be increased (in general, electricity can be produced when prices at market are high, and cooling demand is low).

Direct load control programs are typically operated to control/balance supply and demand at system peak. Focus should be given to direct load control system among consumers. The flexible demand innovative idea requires devices that communicate information, making remote wireless control of appliances located at the average end-user possible, enabling these appliances to dynamically react to changes in the grid status (instead of the usual utilities and very large customers).

Remote controlling of consumption will be done to customers equipped with a separate meter, only for a few hours per day and only for certain equipments eligible to be served (electric boiler, air-condition systems) offering a price discount to the end-users. Peak keeper example (Fig. 7.8) shows an application of direct load control at a region or even at a country level.

The concept requires when enrolled to Peak Keeper program, a device next to the exterior unit of the central air conditioning unit should also be connected. On selected days during the summer months, participating in air-condition (or air-controlled devices) will be automatically coordinated and settled to help manage the demand for electricity. For cooling there will be a limit on the lower temperature and the air-conditioning will not be available to go below that set point for these three months when the system needs it. Knowing the downscaled LST images at 1 km on a quarter basis (real-time satellite data) and having estimated CDD and CDH for evaluating the effect of air temperature on the energy performance of buildings, it gives the option to authorities, urban planners, even the system operator to exclude participants in the program temporarily

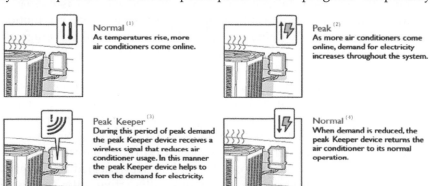

Fig. 7.8. The Peak Keeper concept

allowing them to lower the temperature of the air conditioning unit when needed. The idea for those participating in the load control program is a rate discount or bill credit offered to customers. These rates are usually offered to the utilities' 'largest' industrial and commercial customers. This way, during heat waves, the participants in such programs can have the rate discount benefits and at the same time assist the grid. Through the images utilization a spatial distribution in a city can show exactly in which areas this policy is needed.

There are also established time-of-use rates. Typically two or more periods within a day should be established that correspond to hours when the system load is higher (peak) or lower (off-peak), and charge a higher rate during peak hours. Off-peak hours mostly cover significant parts of the evening and night, as well as weekends. However, since it is about a real time method, the end users can select their commitment and participation in the program with more electric devices (e.g. washing machines, dryers etc) according to price forecasting. Based on the acquired data from satellites (every 15 min), a time series forecasting method can prognose for the next hours the electricity price and the electricity price curve (1 price every 15 min). Therefore, a larger market design that will attract large number of end-users for regulating power can be developed. This can be done via this price-signal for regulating power.

7.3 Exergy Analysis and Summing-up

As a fundamental measure of the thermodynamic deviation of any system in relation to its environment, exergy is equal to the maximum amount of work the system can offer when in thermodynamic equilibrium is with the environment [Ji and Chen, 2005; Hepbasli, 2008; Dincer, 2002; Bejan, 2002; Dincer and Cengel, 2001; Tsatsaronis and Park, 2002]. The expressions of energy (n) and exergy (ψ) efficiencies for the principal types of processes are based on the following. Energy efficiency is defined as:

$$n = \frac{\text{work}}{\text{energy input}} \tag{1}$$

whereas exergy efficiency is defined as:

$$\psi = \frac{\text{work}}{\text{energy input}} \tag{2}$$

$$\psi = \frac{n}{\gamma} \tag{3}$$

When γ is the exergy factor. Therefore, the exergy efficiency is equal to the conventional energy efficiency divided by the exergy factor (Eq. 3). The weighted mean overall exergy efficiency is calculated as:

$$\Psi_{overall} = \sum_{i,k} \frac{n_i}{\gamma_k} \cdot Fr_{i,k} \tag{4}$$

where $\Psi_{overall}$ expresses the weighted mean overall exergy efficiency, n_i stands for the energy efficiency, γ_k is the exergy factor of each energy form used (for instance electricity or diesel oil), and exergy fraction Fr which denotes the rate of the energy form used to the total energy used.

Companies, decision makers, urban planners etc could be based on the theory of exergy analysis implemented all over the city processes and optimize operations, identifying inefficiencies, minimizing losses and ending up to the most needed model ever implemented at a city level.

Spatial planning is a useful tool for planners at a city level. A way more important one is exergy analysis. An integrated systemic analysis could lead to the real available energy analysis within the cities.

7.4 Concluding Remarks

What is clearly necessary is to study the exergy and energy efficiencies related to energy and exergy consumption during the practically endless periods of the city's development. Exergy analysis and its corresponding results offer constructive suggestions for the optimization and improvement of the various industrial subsectors within the urban environment since it provides a linkage between the physical and the engineering world and the surrounding environment. Because of this reason, such an analysis can be utilized in a sustainable sectorial energy planning and management. Spatial planning and its analyses – based on well-known tools -, remote sensing and its numerous applications are among the most powerful ways to be utilized to find out where the energy losses are hidden within the urban environment. Furthermore, via the transportation-energy-food nexus there is room for various activities to be part of this holistic approach and integrated urban systems. Heating systems, vegetables production and EVs (as flexible loads), decentralized electricity production, sustainable water management, and hydrogen-based systems should be among the priorities of urban planners towards the "smart city 2.0" goal.

REFERENCES

Alhamwi, A., Medjroubi, W., Vogt, T. and Agert, C. (2017). GIS-based urban energy systems models and tools: Introducing a model for the optimisation of flexibilisation technologies in urban areas. Applied Energy 191: 1–9.

Bachmaier, A., Narmsara, S., Eggers, J.-B. and Herkel, S. (2016). Spatial distribution of thermal energy storage systems in urban areas connected to district heating for grid balancing—A techno-economical optimization based on a case study. Journal of Energy Storage 8: 349–357.

Bejan, A. (2002). Fundamentals of exergy analysis, entropy generation minimization, and the generation of flow architecture. International Journal of Energy Research 26(7): 545–565.

Ca, V.T., Asaeda, T. and Abu, E.M. (1998). Reductions in air conditioning energy caused by a nearby park. Energy and Buildings 29: 83–92, doi:10.1016/S0378-7788(98)00032-2.

D'Autilia, R. and D'Ambrosi, I. (2015). Is there enough fertile soil to feed a planet of growing cities? Physica A: Statistical Mechanics and its Applications 419: 668–674.

Davies, M., Steadman, P. and Oreszczyn, T. (2008). Strategies for the modification of the urban climate and the consequent impact on building energy use. Energy Policy 36: 4548–4551. doi:10.1016/j.enpol.2008.09.013.

Despommier, D. (2012). Advantages of the vertical farm. pp. 259–275. *In*: Rassia, S.T., Pardalos, P.M. (eds.). Sustainable Environmental Design in Architecture. Springer New York, NY. Available at: http://link.springer.com/10.1007/978-1-4419-0745-5_16 (Accessed: 10 December 2015).

Dincer, I. (2002). The role of exergy in energy policy making. Energy Policy 30(2): 137–149.

Dincer, I. and Cengel, Y.A. (2001). Energy, entropy and exergy concepts and their roles in thermal engineering. Entropy 3(3): 116–149.

Ge, X. and Kremers, E. (2016). Optimization applied with agent based modelling in the context of urban energy planning. Proceedings – Winter Simulation Conference 2016-February, 7408417, pp. 3096–3097.

Goodchild, M.F. (1987). A spatial analytical perspective on GIS. International Journal of Geographical Information Systems 1(4): 327–334.

Gordon, M. (2006). On the frequency of snowfall in metropolitan England. Quarterly Journal of the Royal Meteorological Society 84: 70–72.

Guerreiro, C., de Leeuw, F., Foltescu, V., González Ortiz, A. and Horálek, J. (2015). European Environment Agency, Air quality in Europe: 2015 report. Publications Office, Luxembourg, 2015. http://bookshop.europa.eu/uri? target = EUB:NOTICE:THAL15005:EN:HTML (accessed April 1, 2018).

Haaland, C. and van den Bosch, C.K. (2015). Challenges and strategies for urban green-space planning in cities undergoing densification: A review. Urban Forestry & Urban Greening 14: 760–771.

Helmolt, R. von and Eberle, U. (2007). Fuel cell vehicles: Status 2007. Journal of Power Sources 165(2): 833-843.

Hepbasli, A. (2008). A key review on exergetic analysis and assessment of renewable energy resources for a sustainable future. Renewable and Sustainable Energy Reviews 12(3): 593–661.

Hirano, Y. and Fujita, T. (2012). Evaluation of the impact of the urban heat island on residential and commercial energy consumption in Tokyo. Energy 37: 371–383.

Ji Xi and Chen, G.Q. (2005). Exergy analysis of energy utilization in the transportation sector in China. Energy Policy 34: 1709–1719.

Ji, Y., Xu, P., Ye, Y., Lu, X. and Mao, J. (2016). Energy load superposition and spatial optimization in urban design: A case study, 2016. Computers, Environment and Urban Systems 57: 26–35.

Koroneos, C.J., Nanaki, E.A., and Xydis, G.A. (2012a). Sustainability indicators for the use of resources—The exergy approach. Sustainability 4(8): 1867–1878.

Koroneos, C., Xydis, G. and Polyzakis, A. (2012b). The optimal use of renewable energy sources – The case of Lemnos Island. International Journal of Green Energy 10(8): 860–875, DOI:10.1080/15435075.2012.727929

Kozai, T. (2013). Resource use efficiency of closed plant production system with artificial light: Concept, estimation and application to plant factory. Proceedings of the Japan Academy, Series B 89(10): 447–461. http://dx.doi.org/10.2183/pjab.89.447.

Kozai, T. (2016). Plant production process, floor plan, and layout of PFAL. pp. 203–212. *In*: Takagaki, T.K.N. (ed.). Plant Factory. Academic Press, San Diego.

Kozai, Toyoki, Niu, Genhua, Takagaki and Michiko (eds.). (2015). Plant Factory an Indoor Vertical Farming System for Efficient Quality Food Production, first ed. Academic Press.

Lévay, P.Z., Drossinos, Y. and Thiel, C. (2017). The effect of fiscal incentives on market penetration of electric vehicles: A pairwise comparison of total cost of ownership. Energy Policy 105: 524–533.

Liaros, S., Botsis, K. and Xydis, G. (2016). Technoeconomic evaluation of urban plant factories: The case of Basil (Ocimum basilicum). Science of the Total Environment 554–555: 218–227, doi: 10.1016/j.scitotenv.2016.02.17

Martin, E., Shaheen, S.A., Lipman, T.E. and Lidicker, J.R. (2009). Behavioral response to hydrogen fuel cell vehicles and refueling: Results of California drive clinics. International Journal of Hydrogen Energy 34(20): 8670–8680.

Martos, A., Pacheco-Torres, R., Ordóñez, J. and Jadraque-Gago, E. (2016). Towards successful environmental performance of sustainable cities: Intervening sectors. A review. Renewable and Sustainable Energy Reviews 57: 479–495. doi:10.1016/j.rser.2015.12.095

Millet, P. and Grigoriev, S. (2013). Water electrolysis technologies. pp. 19–41. *In*: Gandía, L.M., Arzamendi, G. and Diéguez, P.M. (eds.). Renewable Hydrogen Technologies. Amsterdam: Elsevier.

Mills, G. (2006). Progress toward sustainable settlements: A role for urban climatology. Theoretical and Applied Climatology 84: 69–76, doi:10.1007/s00704-005-0145-0

Mirzaei, P.A. and Haghighat, F. (2010). Approaches to study Urban Heat Island – Abilities and limitations. Building and Environment 45: 2192–2201.

Moonen, P., Defraeye, T., Dorer, V., Blocken, B. and Carmeliet, J. (2012). Urban physics: Effect of the micro-climate on comfort, health and energy demand. Frontiers of Architectural Research 1: 197–228, doi:10.1016/j.foar.2012.05.002.

Murakami, S., Ooka, R., Mochida, A., Yoshida, S. and Kim, S. (1999). CFD analysis of wind climate from human scale to urban scale. Journal of Wind Engineering and Industrial Aerodynamics 81: 57–81, doi:10.1016/S0167-6105(99)00009-4.

Noorollahi, Y., Yousefi, H. and Mohammadi, M. (2016). Multi-criteria decision support system for wind farm site selection using GIS. Sustainable Energy Technologies and Assessments 13: 38–50.

Panagiotidou, M., Xydis, G. and Koroneos, C. (2016). Environmental siting framework for wind farms: A case study in the Dodecanese Islands. Resources 5(3): 24.

Phillis, Y.A., Kouikoglou, V.S. and Verdugo, C. (2017). Urban sustainability assessment and ranking of cities. Computers, Environment and Urban Systems 64: 254–265.

Power, A. (2001). Social exclusion and urban sprawl: Is the rescue of cities possible? Regional Studies 35: 731–742, doi:10.1080/00343400120084713

Putra, P.A. and Yuliando, H. (2015). Soilless culture system to support water use efficiency and product quality: A review. International Conference on Agro-industry (IcoA): Sustainable and Competitive Agro-industry for Human Welfare Yogyakarta-INDONESIA 2014. Agriculture and Agricultural Science Procedia 3: 283–288. http://dx.doi.org/10.1016/j.aaspro.2015.01.054

Riis, T. and Sandrock, G. (2006). Hydrogen Storage R&D, Priorities and Gaps, in Hydrogen Production and Storage, R&D Priorities and Gaps, Paris, France: IEA Publications.

Rodman, L.C. and Meentemeyer, R.K. (2006). A geographic analysis of wind turbine placement in Northern California. Energy Policy 34(15): 2137–2149.

Sardare, M. and Admane, S. (2013). A review on plant without soil – Hydroponics. International Journal of Research in Engineering and Technology, ISSN: 2319-1163.

Sarralde, J.J., Quinn, D.J., Wiesmann, D. and Steemers, K. (2015). Solar energy and urban morphology: Scenarios for increasing the renewable energy potential of neighborhoods in London. Renewable Energy 73: 10–17.

Seljom, P. and Tomasgard, A. (2015). Short-term uncertainty in long-term energy system models—A case study of wind power in Denmark. Energy Economics 49: 157–167.

Serrao, L., Onori, S., Rizzoni, G. and Guezennec, Y. (2009). A novel model-based algorithm for battery prognosis. International Federation of Automatic Control (IFAC) Proceedings 42(8): 923–928.

The World Bank, Urban population (% of total) | Data, (2018). https://data.worldbank.org/indicator/SP.URB.TOTL.IN.ZS (accessed April 1, 2018).

Toparlar, Y., Blocken, B., Maiheu, B. and van Heijst, G.J.F. (2017). A review on the CFD analysis of urban microclimate. Renewable & Sustainable Energy Reviews 80: 1613–1640.

Tsatsaronis, G. and Park, M.-H. (2002). On avoidable and unavoidable exergy destructions and investment costs in thermal systems. Energy Conversion and Management 43(9–12): 1259–1270.

UNDESA (2014). World's population increasingly urban with more than half living in urban areas | UN DESA | United Nations Department of Economic and Social Affairs (2014). World-urbanization-prospects-2014.html (accessed April 1, 2018).

Union Européenne and Commission Européenne (2015). EU transport in figures 2016. Luxembourg: Publications Office of the European Union.

United Nations Human Settlements Programme, World Health Organization, Kobe Centre (2016). Global report on urban health: equitable, healthier cities for sustainable development. WHO Kobe Centre, Kobe, Japan, 2016. http://

apps.who.int/iris/bitstream/10665/204715/1/9789241565271_eng.pdf (accessed April 1, 2018).

United Nations Population Fund, Urbanization, U.N. Popul. Fund. (2016). Available from: https://www.unfpa.org/urbanization (accessed April 1, 2018).

Van den Berg, M., Wendel-Vos, W., van Poppel, M., Kemper, H., van Mechelen, W. and Maas, J. (2015). Health benefits of green spaces in the living environment: A systematic review of epidemiological studies. Urban Forestry & Urban Greening 14: 806–816.

Vermeulen, T., Merino, L., Knopf-Lenoir, C., Villon, P. and Beckers, B. (2018). Periodic urban models for optimization of passive solar irradiation. Solar Energy 162: 67–77.

Xydis, G. (2012). Effects of air psychrometrics on the exergetic efficiency of a wind farm at a coastal mountainous site – An experimental study. Energy 37: 632–638, DOI: 10.1016/j.energy.2011.10.039

Xydis, G. (2013a). Wind energy to thermal and cold storage—A systems approach. Energy and Buildings 56: 41–47.

Xydis, G. (2013b). The wind chill temperature effect on a large-scale PV plant – An exergy approach. Journal of Progress in Photovoltaics 21(8): 1611–1624, doi: 10.1002/pip.2247.

Xydis, G., Liaros, S. and Botsis, K. (2017). Energy demand analysis via small scale hydroponic systems in suburban areas – An integrated energy-food nexus solution. Science of the Total Environment 593–594: 610–617, doi: 10.1016/j. scitotenv.2017.03.170

Zsuzsa Lévay Petra, Drossinos Yannis and Christian Thiel (2017). The effect of fiscal incentives on market penetration of electric vehicles: A pairwise comparison of total cost of ownership. Energy Policy 105: 524–533.

Index

Index

Color Plate Section

Chapter 1

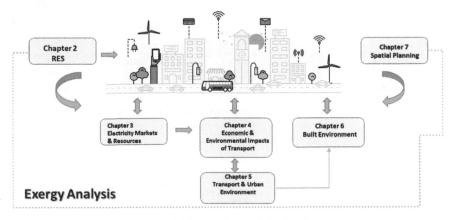

Fig. 1.1. Flow chart of the book

Chapter 4

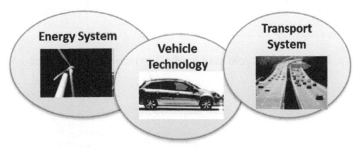

Fig. 4.1. An integrated approach to the economic and environmental assessment of transportation sector within smart cities

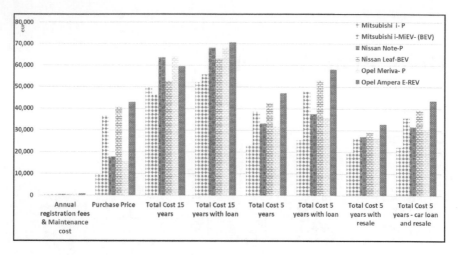

Fig. 4.3. Results of the comparative TCO analysis

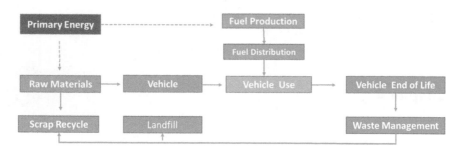

Fig. 4.4. Life cycle stages of alternative fuel vehicles

Chapter 6

Fig. 6.1. World Economic Forum (https://www.weforum.org/) stresses that India faces an unprecedented rate of urbanization

Chapter 7

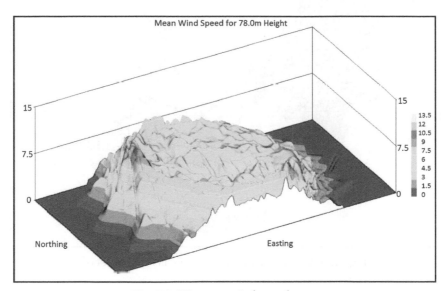

Fig. 7.1. 3D mean wind speed map

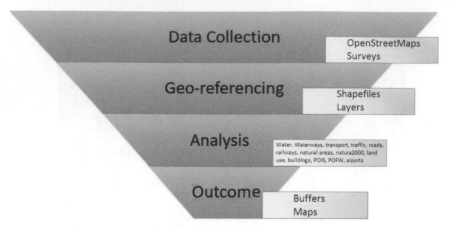

Fig. 7.2. Generic representation of working with GIS tools

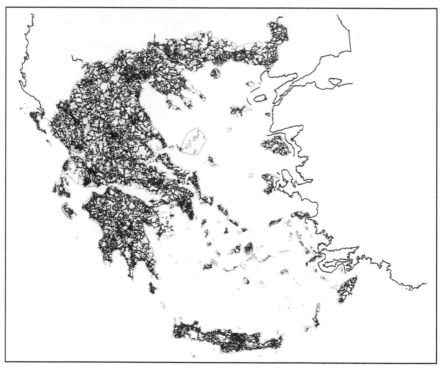

Fig. 7.4. GIS data presented all in one layer for Greece